SIGNAL PROCESSING DESIGN TECHNIQUES

To Joyce, Geoff, and Amber
for their love and support
and to Glenn whose example
taught me to respect
good engineering

SIGNAL PROCESSING DESIGN TECHNIQUES

BRITT RORABAUGH

 TAB Professional and Reference Books

Division of TAB BOOKS Inc.
P.O. Box 40, Blue Ridge Summit, PA 17214

FIRST EDITION

FIRST PRINTING

Copyright © 1986 by TAB BOOKS Inc.

Printed in the United States of America

Reproduction or publication of the content in any manner, without express permission of the publisher, is prohibited. No liability is assumed with respect to the use of the information herein.

Library of Congress Cataloging in Publication Data

Rorabaugh, Britt.
    Signal processing design techniques.

    Bibliography: p.
    Includes index.
    1. Signal processing.   2. Electric filters.
I. Title.
TK5102.5.R57   1986   621.38′043   86-5909
ISBN 0-8306-0457-X

Contents

Preface

This book is an overview of the various aspects of signal processing, but it should not be viewed as all-inclusive. The field is huge and growing—particularly in the digital area. Present applications usually involve a mixture of digital and analog techniques because the speed of digital processors limits the bandwidth of signals on which they can be used. However, since the speed of digital processors is increasing, all-digital techniques are beginning to surpass analog techniques in popularity. Furthermore, the flexibility of digital processing allows the use of new techniques that are impossible to realize with analog techniques.

Introduction

The earliest uses of electricity were substitutes for fire—lightbulbs instead of lanterns and motors instead of steam engines. These purely electrical devices quickly led to electronic devices such as phonographs, radios, amplifiers, and televisions which are primarily concerned with processing of signals. Still further developments led to digital devices which process numbers, data, and perhaps even ideas. The modern field of signal processing spans both analog electronics and digital disciplines. Electrical signals can be processed directly using analog circuits such as op-amp active filters or indirectly by first digitizing the signal into a sequence of numeric values and then processing these values in a digital computing device. While amplifiers and radios are indeed ways to process a signal, the term "signal processing" has come to mean a more limited set of techniques which include primarily frequency selective filtering, spectrum analysis, and spectrum shaping techniques. This book will deal with both analog and digital techniques for performing these types of processing.

Chapter 1 begins with a look at signals and spectra and the mathematics conventionally used to represent them. Chapter 2 continues with a similar treatment of linear systems that are presently used in the great majority of all signal processing applications. Chapter 3 takes the general representation and analysis techniques of Chapter 2 and applies them to frequency selective filters which are a particular type of linear system. Two of the most popular filter types—Butterworth and Chebyshev—are presented in great detail in Chapters 4 and 5. The techniques needed to actually implement one of these filters in analog hardware are presented in Chapter 6.

We make the switch to digital signal processing beginning with Chapter 7, which covers such fundamental concepts as sampling and discrete-time signal and system analysis and forms the foundation for all DSP techniques. Chapter 8 then looks at ways to implement the filters of Chapter 4 or 5 in a digital form, and Chapter 9 examines some digital spectrum analysis techniques that really have no counterpart in the analog world.

Chapter 1

Signals, Spectra, and Noise

T HIS BOOK PRESENTS A COLLECTION OF TECHNIQUES FOR THE
analysis and design of signal processing systems. Such systems can be as simple as a passive resistor-capacitor lowpass filter, or as sophisticated as a dedicated special-purpose computer
for realtime enhancement processing of video signals. Although a
trial-and-error approach may occasionally produce something useful, a few mathematical techniques will prove indispensible in the
design process. Most of these techniques rely on the use of mathematical functions to represent or model real world electronic signals as shown in Fig. 1-1. Actual electronic signals are very
complicated phenomena whose exact behavior may be very difficult or even impossible to describe completely, but simple mathematical models often describe the signals closely enough to produce
very useful results in a variety of practical situations. The distinc-
tion between a signal and its mathematical representation is not
always rigidly observed in signal processing literature—functions
which only *model* signals are commonly referred to as signals and
properties of these models are often presented as properties of the
signals themselves. Purists take warning—this blurring of termi-
nology between a signal and its model crops up everywhere, so you
must learn to live with it.

1

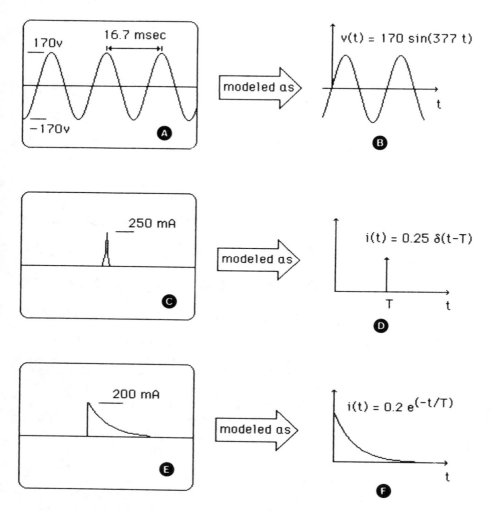

Fig. 1-1. Mathematical models of some practical signals.

This chapter and Chapter 2 are somewhat different from the remainder of the book in that they present theoretical concepts and some fundamental mathematics that are not meant to be used directly for processing actual signals. Instead, this material provides a theoretical basis upon which rest the practical techniques presented in later chapters. Although this material provides valuable insights, a complete understanding of it is not absolutely necessary in order to successfully employ the practical techniques presented in Chapter 3 and beyond.

1.1 SIGNAL MODELS

Mathematical models of signals are generally categorized as either *steady-state* or *transient* models. To understand the difference between these two types, let's examine Fig. 1-2 which shows the typical voltage output from a 1 kHz audio oscillator. This signal exhibits three noticeably different parts—a *turn-on transient* at the beginning, an interval of steady-state operation in the middle, and a *turn-off transient* at the end. We could formulate a single mathematical function to describe all three parts, but for most uses it would be unnecessarily complicated and difficult to work with. In most cases, the primary concern is steady-state behavior, and simplified mathematical representations that ignore the transients are often adequate. The steady-state portion of the oscillator output can be modeled as the sine function shown in Fig. 1-3. Theoretically, this sine function exists for all time, and this might seem to be a contradiction to the obvious fact that the oscillator output only exists for some limited time interval between turn-on and turn-off. However this is really not a problem; over the interval of steady-state operation that we are interested in, the mathematical sine function accurately describes the behavior of the practical oscillator's output voltage. Allowing the mathematical model to assume that the periodic signal exists over all time greatly simplifies matters, since the transients' behavior can be excluded from the model. In situations where the transients are important, they can be modeled

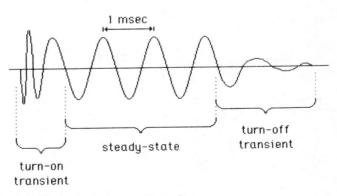

Fig. 1-2. Typical output of an audio oscillator.

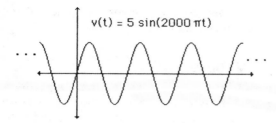

Fig. 1-3. Sine function used to model the steady-state output of the oscillator in Fig. 1-2.

as exponentially saturating and decaying sinusoids as shown in Figs. 1-4 and 1-5. Notice that the amplitude of the saturating exponential envelope continues to increase, but it never quite reaches the steady-state value. Likewise the amplitude of the decaying exponential envelope continues to decrease but it never quite reaches zero. In this context, the steady-state value is sometimes called an *asymptote*, or the envelope can be said to *asymptotically* approach the steady-state value. Of course, such behavior is true only in the pure mathematics of the model—in the real world, signals will eventually get so close to their steady-state values that the difference will be immeasurable.

Steady-state and transient models of signal behavior inherently contradict each other and neither constitutes a "true" description of a particular signal. The selection of an appropriate model requires an understanding of the signal to be modeled and of the implications that a particular choice of model will have for the intended

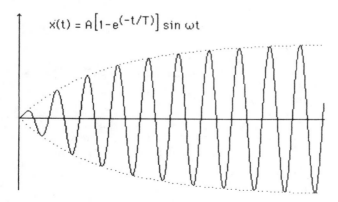

Fig. 1-4. Exponentially saturating sinusoid.

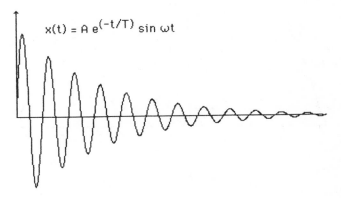

Fig. 1-5. Exponentially decaying sinusoid.

application. The following sections will present details of the more common steady-state and transient signal models.

1.2 STEADY-STATE SIGNALS

Generally, steady-state signals are limited to just sinusoids or sums of sinusoids. This will include virtually any periodic signals of practical interest since such signals can be resolved into sums of weighted and shifted sinusoids using the Fourier analysis techniques presented in Section 1-4. A few basic concepts and definitions will prove useful in work involving steady-state signal models.

1.2.1 Periodicity. Sines, cosines, and squarewaves are all periodic functions The characteristic that makes them periodic is the way in which each of the complete waveforms can be formed by repeating a particular cycle of the waveform over and over at a regular interval as shown in Fig. 1-6. Mathematically, a function $x(t)$ is periodic with a period of T if and only if $x(t+nT) = x(t)$ for all integer values of n.

1.2.2 Symmetry. A function can exhibit a certain symmetry regarding its position relative to the origin. The two major types of symmetry—odd and even—are shown in Fig. 1-7. Symmetry may appear at first to be something that is only "nice-to-know" and not particularly useful in practical applications where the definition of time zero is often somewhat arbitrary. This is far from the case however, because symmetry considerations play an important role in Fourier analysis—especially the discrete Fourier analysis which will be discussed in Chapter 6. Some functions are neither odd or

5

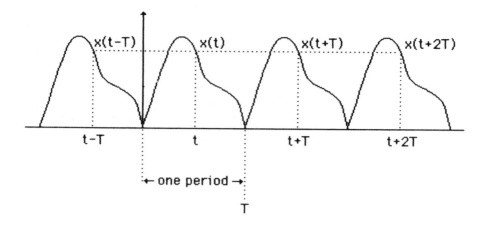

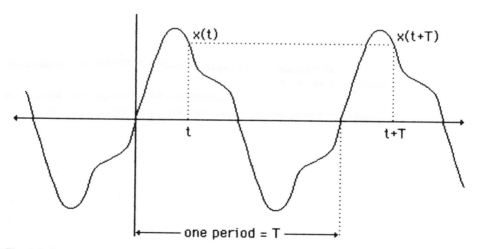

Fig. 1-6. Periodic functions.

even, but Table 1-1 presents formulas that can be used to resolve any periodic function into the sum of an even function and an odd function.

1.3 SINUSOIDS

The sine and cosine functions shown in Fig. 1-8 are together known as *sinusoids*. When a sinusoid is input to a linear system (such

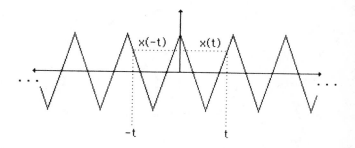

Even symmetry: $x(t) = x(-t)$

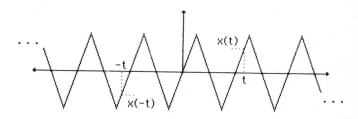

Odd symmetry: $x(t) = -x(-t)$

Fig. 1-7. Symmetry of periodic functions.

Table 1-1. Formulas Concerning Symmetry of Periodic Functions.

$$x(t) = x_{even}(t) + x_{odd}(t)) \qquad \text{(Eq. 1.2-1)}$$

$$x_{even}(t) = \frac{1}{2}\left[x(t) + x(-t) \right] \qquad \text{(Eq. 1.2-2)}$$

$$x_{odd}(t) = \frac{1}{2}\left[x(t) - x(-t) \right] \qquad \text{(Eq. 1.2-3)}$$

even function + even function = even function   (Eq. 1.2-4)

odd function + odd function = odd function   (Eq. 1.2-5)

odd function × odd function = even function   (Eq. 1.2-6)

even function × even function = even function   (Eq. 1.2-7)

even function × odd function = odd function   (Eq. 1.2-8)

7

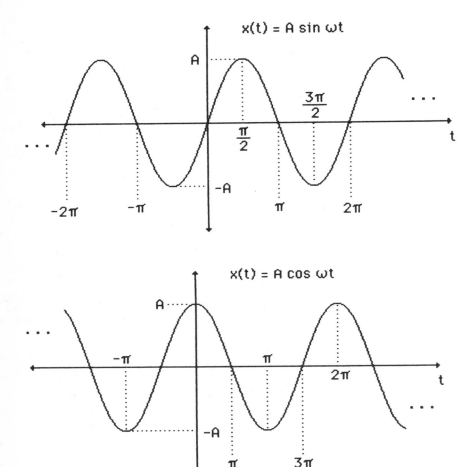

Fig. 1-8. Sine and cosine functions.

as network of resistors, capacitors, and inductors) the output produced will always be a phase-shifted and amplitude-scaled sinusoid of the same frequency. Furthermore, by means of Fourier analysis techniques which we will discuss shortly, any periodic signal can be resolved into a sum of sinusoidal components. Together, these two properties mean that a linear system's output response to any periodic input can easily be calculated by summing the sinusoidal responses due to each of the sinusoidal components of the input.

8

Mathematically, a sinusoidal function x(t) can be defined as:

$$x(t) = A \sin(2\pi t/T) = A \sin(2\pi f t) = A \sin(\omega t)$$

(Eq. 1.3-1)

or

$$x(t) = A \cos(2\pi t/T) = A \cos(2\pi f t) = A \cos(\omega t)$$

(Eq. 1.3-2)

Where

x(t) is the instantaneous amplitude of the sinusoid at time t,

A is the peak amplitude,

T is the period in seconds,

f is the frequency in Hertz, and

ω is the frequency in radians per second.

(Note that the complete argument is in units of radians. This is consistent with the usual convention for the sine and cosine function, but in some situations these functions are defined for arguments in degrees with one period equal to 360°.)

Fig. 1-9. Different forms of sinusoidal functions.

Several different forms of sine and cosine functions are given in Fig. 1-9. Values of these functions for different arguments can easily be computed using a pocket calculator or personal computer. As illustrated in Fig. 1-10, sinusoids can be *phase shifted* left or right by adding respectively either a positive or negative constant to the basic argument of the function. The added constant is called the *phase angle*. A negative phase angle results in a *phase lag*, while a positive angle produces a *phase lead*. Phase shifting leads to several useful equivalences between various sine and cosine func-

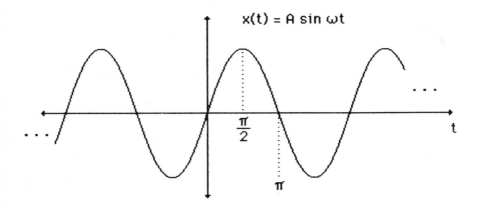

$$x(t) = A \sin \omega t$$

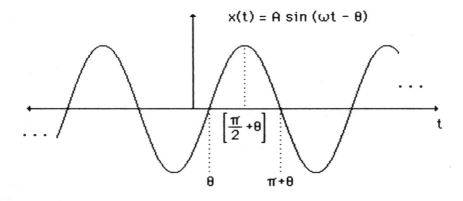

$$x(t) = A \sin (\omega t - \theta)$$

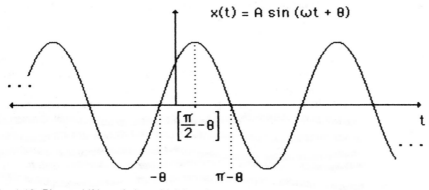

$$x(t) = A \sin (\omega t + \theta)$$

Fig. 1-10. Phase shifting of sinusoidal functions.

tions. These equivalences and some additional *trigonometric identities* are listed in Tables 1-2 and 1-3.

1.4 FREQUENCY SPECTRA OF PERIODIC SIGNALS

As mentioned earlier, periodic functions can be resolved into summations of weighted and shifted sinusoids. The *Fourier series*, shown in Fig. 1-11, is a mathematical technique for determining the relative strengths and phasing of the sinusoidal components in any particular periodic signal. Examination of this summation shows that it contains only a dc component plus sinusoids whose frequencies are integer multiples of the original function's *fundamental* frequency. The fundamental frequency is f_o, while $2f_o$ is the *second harmonic*, $3f_o$ is the *third harmonic*, and so on forever. Here is one of the differences between theory and practice. Theory, in the form of Fourier analysis, says that in general, signals will contain an infinite number of sinusoidal harmonic components. However, in the real world, practical periodic signals will contain only a finite number of observable harmonics. Removing some of the higher harmonics will, of course, distort the shape of the signal to some degree. Figure 1-12 shows how the shape of a squarewave becomes distorted as increasing numbers of harmonic frequencies are removed. In all signals of practical interest the relative amplitude of the harmonic components tends to decrease strongly at the higher frequencies. Therefore, a signal can usually be modeled with an acceptably low amount of distortion by using a reasonably small number of lower-frequency harmonic components. In the case of Fig. 1-12, we see that a reasonable approximation to a squarewave

Table 1-2. Equivalences Between Phase-Shifted Sinusoids.

$$\cos t = \sin (t + \pi/2) \qquad\qquad (Eq. 1.3-3)$$

$$\cos t = \cos (t + 2\pi n), \quad n = \text{any integer} \qquad (Eq. 1.3-4)$$

$$\sin t = \sin (t + 2\pi n), \quad n = \text{any integer} \qquad (Eq. 1.3-5)$$

$$\sin t = \cos (t - \pi/2) \qquad\qquad (Eq. 1.3-6)$$

$$\cos t = -\cos (t + (2n + 1)\pi), \quad n = \text{any integer} \qquad (Eq. 1.3-7)$$

$$\sin t = -\sin (t + (2n + 1)\pi), \quad n = \text{any integer} \qquad (Eq. 1.3-8)$$

Table 1-3. Trigonometric Identities.

$$\tan x = \frac{\sin x}{\cos x} \qquad \text{(Eq.1.3-9)}$$

$$\sin(-x) = -\sin x \qquad \text{(Eq.1.3-10)}$$

$$\cos(-x) = \cos x \qquad \text{(Eq.1.3-11)}$$

$$\tan(-x) = -\tan x \qquad \text{(Eq.1.3-12)}$$

$$(\cos x)^2 + (\sin x)^2 = 1 \qquad \text{(Eq.1.3-13)}$$

$$\sin(x+y) = (\sin x)(\cos y) + (\cos x)(\sin y) \qquad \text{(Eq.1.3-14)}$$

$$\cos(x+y) = (\cos x)(\cos y) - (\sin x)(\sin y) \qquad \text{(Eq.1.3-15)}$$

$$\tan(x+y) = \frac{(\tan x) + (\tan y)}{1 - (\tan x)(\tan y)} \qquad \text{(Eq.1.3-16)}$$

$$\sin(2x) = 2(\sin x)(\cos x) \qquad \text{(Eq.1.3-17)}$$

$$\cos(2x) = (\cos x)^2 - (\sin x)^2 \qquad \text{(Eq.1.3-18)}$$

$$\tan(2x) = \frac{2(\tan x)}{1 - (\tan x)^2} \qquad \text{(Eq.1.3-19)}$$

$$(\sin x)(\sin y) = \frac{1}{2}[-\cos(x+y) + \cos(x-y)] \qquad \text{(Eq.1.3-20)}$$

$$(\cos x)(\cos y) = \frac{1}{2}[\cos(x+y) + \cos(x-y)] \qquad \text{(Eq.1.3-21)}$$

$$(\sin x)(\cos y) = \frac{1}{2}[\sin(x+y) + \sin(x-y)] \qquad \text{(Eq.1.3-22)}$$

$$(\sin x) + (\sin y) = 2\sin\frac{x+y}{2}\cos\frac{x-y}{2} \qquad \text{(Eq.1.3-23)}$$

$$(\cos x) + (\cos y) = 2\cos\frac{x+y}{2}\cos\frac{x-y}{2} \qquad \text{(Eq.1.3-24)}$$

can be obtained by including only ten harmonics in the Fourier series summation.

1.4.1 Other Forms of the Fourier Series. The trigonometric form of the Fourier series shown in Fig. 1-11 makes it easy to visualize periodic signals as summations of sines and cosines, but in many cases involving substantial mathematical manipulation,

The trigonometric form of the Fourier series is given by:

$$x(t) = \frac{a_0}{2} + \sum_{n=1}^{\infty} (a_n \cos(n\omega_0 t) + (b_n \sin(n\omega_0 t)))$$

(Eq. 1.4-1)

$$a_0 = \frac{2}{T} \int_{-T/2}^{T/2} x(t)\, dt$$

(Eq. 1.4-2)

$$a_n = \frac{2}{T} \int_{-T/2}^{T/2} x(t) \cos(n\omega_0 t)\, dt$$

(Eq. 1.4-3)

$$b_n = \frac{2}{T} \int_{-T/2}^{T/2} x(t) \sin(n\omega_0 t)\, dt$$

(Eq. 1.4-4)

where

T = period of $x(t)$

$\omega_0 = \frac{2\pi}{T}$ = fundamental radian frequency of $x(t)$

$= 2\pi f_0$

Since n can assume only integer values, x(t) will be expressed as a sum of sinusoids having frequencies of $\omega_0, 2\omega_0, 3\omega_0 \ldots$ In general, there will also be a dc component of $a_0/2$.

Fig. 1-11. Trigonometric form of the Fourier series.

Fundamental plus all harmonics

$$x(t) = A \sum_{n=0}^{\infty} \left[\frac{\sin((2n+1)\omega t)}{2n+1} \right]$$

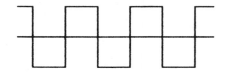

Fundamental plus ten harmonics

$$x(t) = A \sum_{n=0}^{10} \left[\frac{\sin((2n+1)\omega t)}{2n+1} \right]$$

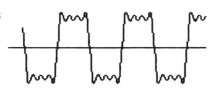

Fundamental plus three harmonics

$$A \left(\sin \omega t + \frac{1}{3} \sin 3\omega t \right.$$
$$\left. + \frac{1}{5} \sin 5\omega t + \frac{1}{7} \sin 7\omega t \right)$$

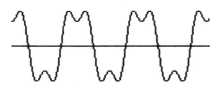

Fundamental plus one harmonic

$$A \left(\sin \omega t + \frac{1}{3} \sin 3\omega t \right)$$

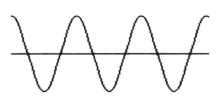

Fundamental only

$$A \sin \omega t$$

Fig. 1-12. Squarewave distorted by removal of higher harmonic components.

the spectrum of a signal is much more useful in an exponential form as shown in Fig. 1-13. This form of the series is easily derived from the trigonometric form by using Euler's identity which is given in Fig. 1-14. Sometimes, the characteristics of a signal's spectrum are hard to visualize from the Fourier series formula. Therefore, the spectrum of a periodic signal is often presented graphically as a *magnitude spectrum* with an accompanying *phase spectrum* as shown in Fig. 1-15.

     1.4.2 Dirichlet Conditions. Although the Fourier series

The exponential form of the Fourier series is given by:

$$x(t) = \sum_{n=-\infty}^{\infty} c_n \, e^{j2\pi n f_0 t} \qquad \text{(Eq. 1.4-5)}$$

where

$$c_n = \frac{1}{T} \int_T x(t) \, e^{-j2\pi n f_0 t} \, dt \qquad \text{(Eq. 1.4-6)}$$

In general, the values of c_n are complex. The real and imaginary parts of c_n are denoted as $\mathrm{Re}\{c_n\}$ and $\mathrm{Im}\{c_n\}$, respectively. Often it is more convenient to express the complex values of c_n as a magnitude $|c_n|$ and a phase angle θ_n:

$$|c_n| = \sqrt{(\mathrm{Re}\{c_n\})^2 + (\mathrm{Im}\{c_n\})^2} \qquad \text{(Eq. 1.4-7)}$$

$$\theta_n = \tan^{-1}\left[\frac{\mathrm{Im}\{c_n\}}{\mathrm{Re}\{c_n\}} \right] \qquad \text{(Eq. 1.4-8)}$$

Fig. 1-13. Exponential form of the Fourier series.

Euler's identity is useful for relating sinusoids and exponentials:

$$e^{jx} = \cos x + j \sin x \qquad \text{(Eq. 1.4-9)}$$

$$e^{-jx} = \cos x - j \sin x \qquad \text{(Eq. 1.4-10)}$$

An equivalent form is sometimes more convenient:

$$\cos x = \frac{e^{jx} + e^{-jx}}{2} \qquad \text{(Eq. 1.4-11)}$$

$$\sin x = \frac{e^{jx} - e^{-jx}}{2j} \qquad \text{(Eq. 1.4-12)}$$

Fig. 1-14. Euler's identity.

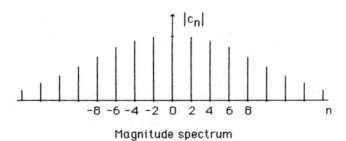

Magnitude spectrum

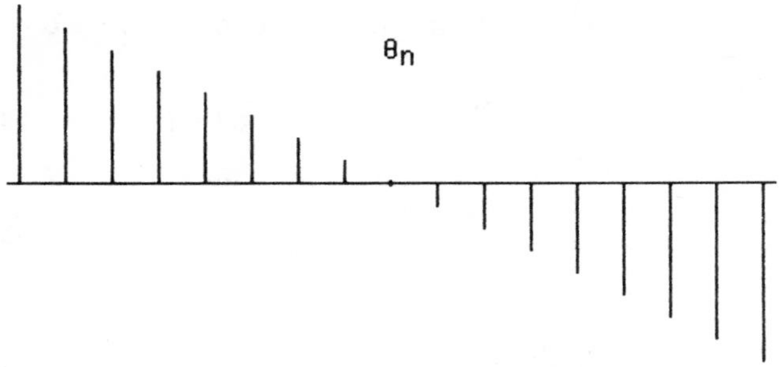

Phase spectrum

Fig. 1-15. Line spectrum of a periodic signal.

16

can be applied to virtually all period signals of practical interest, some functions do exist for which the series will not provide a meaningful representation of the original function. For the benefit of math lovers and masochists (is there a difference?) Fig. 1-16 presents the Dirichlet conditions that must be satisfied to guarantee convergence of the series for any particular function.

1.5 TRANSIENT SIGNALS

There are several transient signals, modeled by aperiodic functions, which are frequently encountered in signal processing work. These include the unit step, unit impulse, and exponentials which are each discussed in the following sections.

1.5.1 Unit Step. Consider the circuit shown in Fig. 1-17. When the switch is open, the voltage between terminals A and B is zero. When the switch is closed, the voltage between A and B will suddenly jump from zero to nine volts. An abrupt change in

---

For the Fourier series coefficients to exist, x(t) must be absolutely integrable over a period:

$$\int_T |x(t)|\, dt \;<\; \infty \qquad\qquad \text{(Eq. 1.4-13)}$$

For the Fourier series to converge uniformly, x(t) must be finite and have only a finite number of discontinuities in each period. These conditions are often called the Dirichlet conditions.

Often it is sufficient for the Fourier series to be convergent in the mean, rather than uniformly convergent. In this case, x(t) need only be integrable square over a period:

$$\int_T |x(t)|^2\, dt \;<\; \infty \qquad\qquad \text{(Eq. 1.4-14)}$$

Fig. 1-16. Dirichlet conditions.

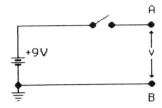

Fig. 1-17. Circuit used in discussion of the unit step function.

level such as this is represented mathematically as a *step function*. Figure 1-18 shows a *unit step* which shifts its level from zero to one at time zero. This function can be multiplied by a constant gain factor and time shifted in order to represent virtually any abrupt dc level shift of practical interest. In most signal processing and linear systems literature the unit step is usually denoted as $u_1(t)$.

1.5.2 Unit Impulse. When a switch is opened or closed in a circuit containing reactive components (i.e., capacitors or inductors), a spike of voltage or current as shown in Fig. 1-1C may be produced. Although this spike has a finite amplitude and a non-zero risetime and falltime, it is often convenient to represent it mathematically as an impulse having zero width and infinite amplitude. Although multiplying zero by infinity is sometimes a questionable operation, in the case of impulse functions the zero width can be multiplied by the infinite amplitude to yield a finite area for the impulse. (Mathematicians will shudder at this explanation, as they use the theory of distributions to explain impulses. However, for our purposes the bottom line is the same regardless of the route taken to get there. If this sounds like gibberish, don't worry about it—the successful selection and application of signal processing techniques does not usually require an intuitive understanding of impulse functions.) A *unit impulse* occurs at time zero and has an area of one. As with the unit step, the unit impulse can be time shifted and multiplied by a constant gain in order to represent almost any

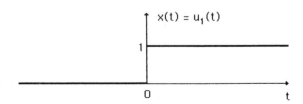

Fig. 1-18. Unit step function.

spiking phenomenon of practical interest. The unit impulse is also called the *Dirac delta function* and it is denoted as either $u_o(t)$ or $\delta(t)$. The special relationships between the unit impulse, unit step, and two other elementary time functions are presented in Fig. 1-19.

1.5.3 Decaying Exponential. Consider the circuit shown in Fig. 1-20. The battery is initially connected across the capacitor and the voltage across the capacitor will be six volts. If the switch is moved from position A to position B, the voltage across the capacitor terminals will decay as shown. Mathematically, the shape of this decaying waveform is described as a *decaying exponential* and is shown in Fig. 1-21. As time increases the amplitude of the function approaches closer and closer to (but never quite reaches) zero. The function asymptotically approaches zero and the horizontal line at $y = 0$ is an asymptote.

1.5.4 Saturating Exponential. Referring again to the circuit of Fig. 1-20, let's assume that the switch is in position B and the voltage across the capacitor is zero. If the switch is moved to position A, the voltage across the capacitor will begin to increase as a *saturating exponential* whose general form is shown in Fig. 1-22. The function approaches an asymptote at $y = \beta$.

1.6 FOURIER TRANSFORM

Fourier series analysis works only with periodic signals and cannot be used to obtain the spectrum of an aperiodic signal. Instead we must use a related technique known as the Fourier transform. Just as with the Fourier series, the transform has both an exponential and trigonometric form which are shown in Fig. 1-23. Rather than continually write out these integrals, the shorthand notation of Fig. 1-24 is commonly used. The values of the frequency function $X(f)$ are generally complex, and their magnitude and phase components are often plotted as shown in Fig. 1-25.

Consider the case of an entire periodic signal and an aperiodic signal corresponding to just one of the periods as shown in Fig. 1-26. As depicted in the figure, a special relationship exists between the spectra of these two signals. The Fourier transform of the aperiodic signal is a continuous frequency function which is equivalent to the envelope of the line spectrum produced by the Fourier series. Usually a function $x(t)$ must be absolutely integrable (Eq. 1.4-13) to guarantee the existence of its transform $X(f)$. However there are some functions that do not satisfy Equation 1.4-13 yet still have valid Fourier transforms.

Unit ramp

$u_2 = 0, \quad t < 0$
$\quad = t, \quad t \geq 0$

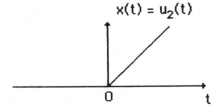

$$u_2(t) = \int u_1(t) \, dt \qquad u_1(t) = \frac{d}{dt} u_2(t)$$

Unit step

$u_1(t) = 0, \quad t < 0$
$\quad\quad = t, \quad t \geq 0$

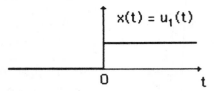

$$u_1(t) = \int u_0(t) \, dt \qquad u_0(t) = \frac{d}{dt} u_1(t)$$

Unit impulse

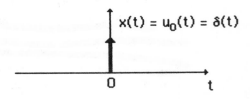

$$u_0(t) = \int u_{-1}(t) \, dt \qquad u_{-1}(t) = \frac{d}{dt} u_0(t)$$

Unit doublet

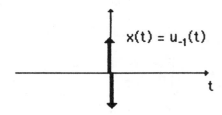

Fig. 1-19. Elementary time functions.

20

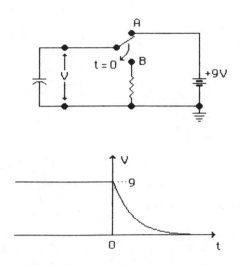

Fig. 1-20. Circuit used in discussion of decaying and saturating exponentials.

1.7 FOURIER TRANSFORMS-IN-THE-LIMIT

As we have seen, the spectra of periodic functions are obtained with Fourier series, while the spectra of aperiodic timelimited functions are obtained with Fourier transforms. In many practical situations there is a requirement to analyze signals that have both periodic and aperiodic components. Mixing Fourier transforms and series' in such an analysis can quickly produce an ugly mess. To make such analysis more convenient, the spectra of most periodic functions can be obtained by performing limiting operations upon

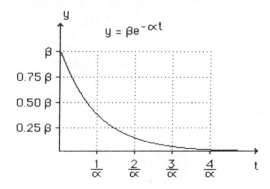

Fig. 1-21. Decaying exponential.

21

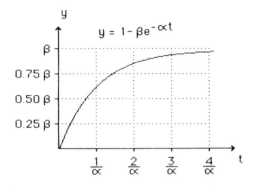

Fig. 1-22. Saturating exponential.

functions having established transforms. When the spectrum of a periodic function is obtained via a Fourier series, the spectrum will consist of lines located at the fundamental frequency and its harmonics. When the spectrum of this same function is obtained via Fourier transforms-in-the-limit, it will consist of weighted impulses occurring at the fundamental frequency and its harmonics. Obviously these two different mathematical representations must be equivalent in their physical significance.

1.8 PRACTICAL USE OF FOURIER ANALYSIS TECHNIQUES

The formal definitions of the Fourier series and transform are important in academic and advanced theoretical work, and they do involve some integral calculus. In most practical analysis work, direct use of the calculus can be avoided by using tables of well-established Fourier transform pairs which relate common signals to their spectra. In addition, a number of well-established properties can be used to manipulate the information obtained from these tables in order to modify it to fit a particular application. Table 1-4 lists a number of useful Fourier transform pairs and Table 1-5 lists some properties that can be used to adapt these pairs to fit various situations.

1.9 NOISE

Noise contaminates virtually all signals to some degree, and is best described as a random process that does not meet the requirements for application of the Fourier transform. However, us-

The exponential form of the Fourier transform is given as:

$$X(f) = \int_{-\infty}^{\infty} x(t)\ e^{-j2\pi ft}\ dt \qquad \text{(Eq. 1.6-1)}$$

The exponetial form of the inverse Fourier transform is given by:

$$x(t) = \int_{-\infty}^{\infty} X(f)\ e^{j2\pi ft}\ df \qquad \text{(Eq. 1.6-2)}$$

The Fourier transform is occassionally seen in trigonometric form which is given by:

$$X(f) = \int_{-\infty}^{\infty} x(t)\ \cos(2\pi ft)\ dt\ -\ j\int_{-\infty}^{\infty} x(t)\ \sin(2\pi ft)\ dt \qquad \text{(Eq. 1.6-3)}$$

$$x(t) = \int_{-\infty}^{\infty} X(f)\ \cos(2\pi ft)\ df\ +\ j\int_{-\infty}^{\infty} X(f)\ \sin(2\pi ft)\ df \qquad \text{(Eq. 1.6-4)}$$

Fig. 1-23. Fourier transform.

In mathematical notation,

$$X(\omega) = \mathcal{F}[\, x(t)\,] \qquad\qquad (Eq.\ 1.6\text{-}5)$$

means that $X(\omega)$ is the Fourier transform of $x(t)$.

Likewise,

$$x(t) = \mathcal{F}^{-1}[\, X(\omega)\,] \qquad\qquad (Eq.\ 1.6\text{-}6)$$

means that $x(t)$ is the inverse Fourier transform of $X(\omega)$.

Fig. 1-24. Shorthand notation for the Fourier transform integrals.

ing statistical techniques, the power spectrum of many common noise processes can be obtained. A great many noise sources encountered in the real world can be adequately modeled as Gaussian processes with flat spectra. The power of such signals will be distributed more or less evenly over all frequencies of practical interest. Usually the signals that we want to process will have spectra that are confined to a reasonably small frequency range. Thus, we can remove most of the noise contamination from our signals of interest by using filters which severely attenuate all frequencies outside the range occupied by the signal. This will remove most of the noise. Of course, that part of the noise which occupies the same frequencies as the signal cannot be removed with such a filter, but the improvement due to removing the rest is usually good enough for a great many applications.

$$|X(\omega)| = \sqrt{[\,Re\{\,X(\omega)\}\,]^2 + [\,Im\{\,X(\omega)\}\,]^2}$$

(Eq. 1.6-7)

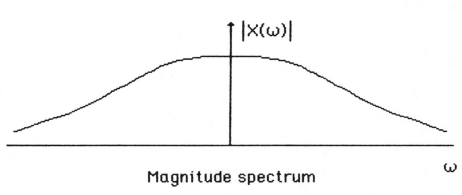

Magnitude spectrum

$$arg\{\,X(\omega)\,\} = tan^{-1}\left[\frac{Im\{\,X(\omega)\,\}}{Re\{\,X(\omega)\,\}}\right]$$ (Eq. 1.6-8)

Phase spectrum

Fig. 1-25. Magnitude and phase plots of a complex-valued function of frequency.

25

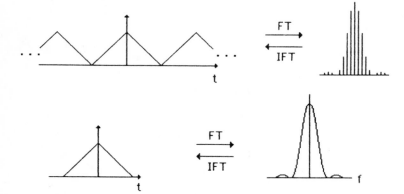

Fig. 1-26. Fourier transforms of an entire periodic function and of a single period.

Table 1-4. Fourier Transform Pairs.

$x(t)$	$X(\omega)$	
1	$2\pi\,\delta(\omega)$	(Eq. 1.8-1)
$u_1(t)$	$\dfrac{1}{j\omega} + \pi\,\delta(\omega)$	(Eq. 1.8-2)
$\delta(t)$	1	(Eq. 1.8-3)
t	$2\pi\,j\,\delta'(\omega)$	(Eq. 1.8-4)
t^n	$2\pi\,j^n\,\delta^{(n)}(\omega)$	(Eq. 1.8-5)
$\sin\omega_0 t$	$-j\pi\big(\delta(\omega-\omega_0) - \delta(\omega+\omega_0)\big)$	(Eq. 1.8-6)
$\cos\omega_0 t$	$\pi\big(\delta(\omega-\omega_0) + \delta(\omega+\omega_0)\big)$	(Eq. 1.8-7)
$e^{-at}\,u_1(t)$	$\dfrac{1}{j\omega + a}$	(Eq. 1.8-8)
$u_1(t)\,e^{-at}\sin\omega_0 t$	$\dfrac{\omega_0}{(a+j\omega)^2 + \omega_0^2}$	(Eq. 1.8-9)
$u_1(t)\,e^{-at}\cos\omega_0 t$	$\dfrac{a+j\omega}{(a+j\omega)^2 + \omega_0^2}$	(Eq. 1.8-10)

Table 1-5. Fourier Transform Properties.

Time function	Transform	
$x(t)$	$X(f)$	(Eq. 1.8-11)
$a\,x(t)$	$a\,X(f)$	(Eq. 1.8-12)
$y(t) + x(t)$	$Y(f) + X(f)$	(Eq. 1.8-13)
$\dfrac{d}{dt}\,x(t)$	$j\omega\,X(f)$	(Eq. 1.8-14)
$\displaystyle\int_{-\infty}^{t} x(\tau)\,d\tau$	$\dfrac{X(f)}{j\omega}$	(Eq. 1.8-15)
$e^{-j2\pi f_0 t}\,x(t)$	$X(f+f_0)$	(Eq. 1.8-16)
$x(t-\tau)$	$e^{-j\omega\tau}\,X(f)$	(Eq. 1.8-17)
$\displaystyle\int_{-\infty}^{\infty} h(t-\tau)\,x(\tau)\,d\tau$	$H(f)\,X(f)$	(Eq. 1.8-18)
$x\left(\dfrac{t}{a}\right)$	$a\,X(af)$	(Eq. 1.8-19)

Chapter 2

Systems, Circuits, and Networks

<hr>

S YSTEM THEORY IS USUALLY DIVIDED INTO THE TWO SEP-
arate topics of analysis and synthesis. System analysis is used
to estimate the input/output behavior of a given system, while synthesis is used to design a system that will exhibit a specific desired
relationship between its input and output. Although synthesis appears to have more practical uses, a good grasp of network analysis techniques is usually required to understand and apply the
techniques of network synthesis. In this chapter I will concentrate
primarily on continuous-time linear time-invariant (CTLTI) systems
since they play a fundamental role in basic signal processing.

2.1 SYSTEMS

Within the context of signal processing, a *system* is something
which accepts one or more input signals and operates upon them
to produce one or more output signals. Filters, amplifiers, and
digitizers are some of the systems used in various signal processing applications. As we saw in Chapter 1, real world signals are
conveniently represented as mathematical functions, permitting
them to be analyzed with powerful mathematical techniques. In a
similar fashion, systems are often represented mathematically as

A system H having an input x(t) and output y(t) can be represented as:

$$y(t) = H[x(t)] \qquad\qquad (Eq. 2.1-1)$$

H is homogeneous if

$$H[a\, x(t)] = a\, H[x(t)] \qquad\qquad (Eq. 2.1-2)$$

H is additive if

$$H[x(t) + f(t)] = H[x(t)] + H[f(t)] \qquad\qquad (Eq. 2.1-3)$$

H exhibits superposition and is linear if

$$H[a\, x(t) + b\, f(t)] = a\, H[x(t)] + b\, H[f(t)] \qquad\qquad (Eq. 2.1-4)$$

H is time-invariant if

$$y(t-\tau) = H[x(t-\tau)] \qquad\qquad (Eq. 2.1-5)$$

H is causal if

$$H[x(t)] = H[f(t)] \quad \text{for } t \le t_0 \qquad\qquad (Eq. 2.1-6)$$

given that

$$x(t) = f(t) \quad \text{for } t \le t_0$$

Fig. 2-1. Mathematical properties of systems.

operators, which operate on input functions to produce output functions. A system H, having an input of x(t) and producing an output of y(t) can be represented as shown in Eq. 2.1-1. If x and y are functions of continuous time (representing analog input and output signals respectively), the system H is a *continuous-time* system.

2.1.1 Linearity. As illustrated in Fig. 2-2, a system H is *homogeneous* if multiplying the input signal, x(t), by a constant gain, a, has no effect other than to multiply the output function, y(t), by the same constant gain. Thus, if Equation 2.1-2 is satisfied, H is homogeneous. The system H exhibits *additivity*, depicted in Fig. 2-3, if the output produced for the sum of two input signals is equal to the sum of the outputs produced for each input individually. Thus, if Equation 2.1-3 is satisfied, H is additive. Acting together, homogeneity and additivity produce the property of *superposition*. If a system exhibits superposition, then it is a linear system. Thus, if Equation 2.1-4 is satisfied, H is linear.

2.1.2 Time-invariance. A system whose characteristics do not change over time is called a *time-invariant* system. Therefore, a relaxed time-invariant system will produce the same output from a given input regardless of when the input is applied. (A *relaxed* system is one which is still not responding to any previously applied input.) Mathematically, a system H is time-invariant if and only if Equation 2.1-5 is satisfied. Sometimes a time-invariant system is also referred to as *stationary* or *fixed*. Conversely, a system which is not time-invariant is called *time-varying, variable,* or *nonstationary*.

If H is homogeneous, circuits A and B are equivalent.

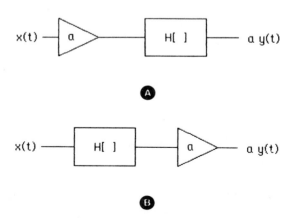

Fig. 2-2. Homogeneous system.

If H is additive, circuits A and B are equivalent.

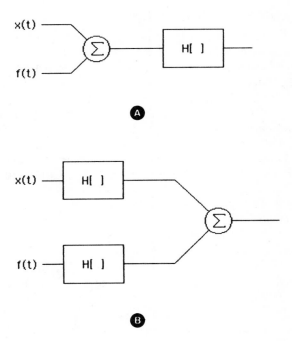

Fig. 2-3. Additive system.

In analog signal processing, continuous-time linear time-invariant systems are encountered so frequently that they are usually referred to as CTLTI systems. All of the systems and functions presented in this book will be both linear and time-invariant (but not necessarily continuous time).

2.1.3 Causality. Another property of concern in signal processing systems is *causality*. Put simply, a causal system "doesn't laugh until its' tickled." More precisely, a causal system is one in which the present output never depends upon future values of the input. Mathematically, a system H is causal if and only if Equation 2.1-6 is satisfied. *Noncausal* or *anticipatory* systems are abundant in theory, but they cannot exist in the real world. (However, a causal realization is possible for a noncausal system in which the present output depends at most on past, present and a finite extent of future input. All that needs to be done is to delay the output for a finite interval until all the required future input values have entered the system and are available for contributions to the output.

2.2 REPRESENTING SYSTEMS

Differential equations are one of the most general ways to express the relationship between the inputs and outputs of both linear and nonlinear systems. The input x(t) and output y(t) of a CTLTI system can be related by a differential equation of the form shown in Fig. 2-4. In all but the simplest cases, direct solution of differential equations can be difficult or even impossible. Fortunately, several more convenient methods such as impulse response, step response, and transfer functions are available for characterizing CTLTI systems. These alternative descriptions of systems are each described in the following sections, and the relationships between them are summarized in Fig. 2-5. The intent in this chapter is to just introduce these concepts—their practical uses will be discussed in Chapter 3.

2.2.1 Impulse Response. A system's impulse response, often denoted as h(t), is the output response produced when a unit impulse, $u_o(t)$ or $\delta(t)$, is applied to the input of the previously relaxed system. (For a discussion of the unit impulse refer to Section 1.5.2) Figure 2-6 shows the impulse response of a simple low-pass filter. The impulse response is a particularly useful description of a system since the output, y(t), due to any input, x(t), can be computed for a causal linear time-invariant system H by just convolving the input function with the system's impulse response using the convolution integral defined in Equation 2.2-2. Once again calculus rears its ugly head! However, direct use of the convolution integral is rarely necessary in practice. As we saw in the discussion of Fourier transforms and will soon reaffirm for Laplace transforms, convolution in the time domain is equivalent to multiplication

$$b_0 y + b_1 \frac{d}{dt} y + b_2 \frac{d^2}{dt^2} y + \ldots + b_m \frac{d^m}{dt^m} y$$

$$= a_0 x + a_1 \frac{d}{dt} x + a_2 \frac{d^2}{dt^2} x + \ldots + a_m \frac{d^m}{dt^m} x$$

(Eq. 2.2-1)

Fig. 2-4. Differential equation for a CTLTI system.

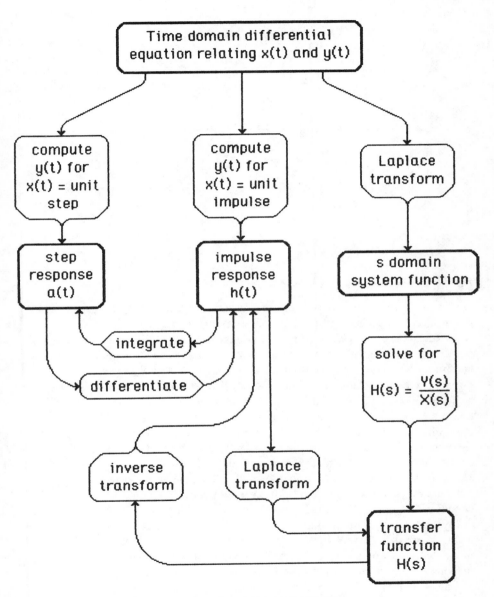

Fig. 2-5. Relationships between various linear system descriptions.

in the frequency domain. Therefore, a convolution of two functions can be performed indirectly by taking the transforms of the functions, multiplying together the two resulting spectra, and then in-

33

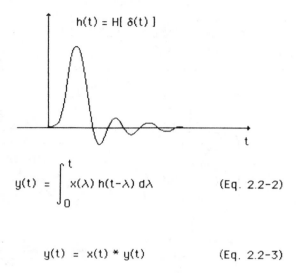

$h(t) = H[\ \delta(t)\ ]$

$$y(t) = \int_0^t x(\lambda)\, h(t-\lambda)\, d\lambda \qquad \text{(Eq. 2.2-2)}$$

$$y(t) = x(t) * y(t) \qquad \text{(Eq. 2.2-3)}$$

Fig. 2-6. Impulse response of a typical lowpass filter.

verse transforming the product. To avoid writing out a cumbersome integral, convolution is often abbreviated as in Equation 2.2-3.

2.2.2 Step Response. In some situations it may be advantageous to analyze the behavior of liner systems by using their response to a unit step rather than a unit impulse. A digital pulse stream can be modeled as a weighted and shifted sum of unit steps as shown in Fig. 2-7. Since this sum is a *linear* combination of unit steps, a CTLTI systems's response to such a signal can be determined by forming a similarly weighted and shifted sum of the response to a single unit step. This response to a single unit step is of course the *step response* and is usually denoted as a(t).

The step response of a typical lowpass filter is shown in Fig. 2-8. Such a response will often exhibit overshoot and ringing as shown in the figure, but depending upon the damping present in the system these phenomena may be minimized or eliminated. This will be discussed further in the section on damping and stability. Since the unit impulse is the derivative of a unit step, the impulse response of a system is equal to the derivative of its step response as stated in Equation 2.2-4; or conversely, the step response is equal to the integral of the impulse response as stated in Equation 2.2-5.

2.3 LAPLACE TRANSFORM TECHNIQUES

As mentioned previously, direct solution of differential equa-

34

This digital pulse signal:

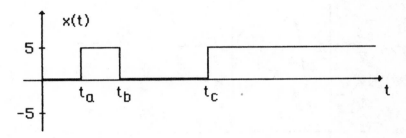

can be resolved into a summation of the following
weighted and shifted unit steps:

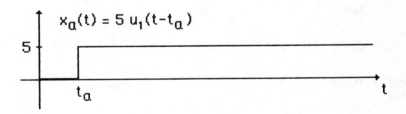

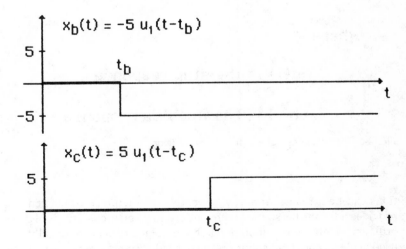

Fig. 2-7. Digital pulse stream resolved into a weighted and shifted sum of unit steps.

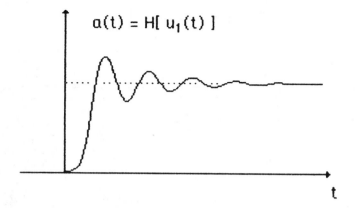

$$\frac{d}{dt}\ a(t)\ =\ h(t) \qquad\qquad (Eq.\ 2.2\text{-}4)$$

$$a(t)\ =\ \int_{0}^{t} h(\tau)\ d\tau \qquad\qquad (Eq.\ 2.2\text{-}5)$$

where

a(t) is the step response

h(t) is the impulse response

Fig. 2-8. Step response of a typical lowpass filter.

tions can be a real nuisance. Fortunately, the Laplace transform provides a means for solving many differential equations using only ordinary algebraic manipulations. The Laplace transform usually used in electronics applications is the one-sided transform presented in Fig. 2-9. If both sides of a differential equation are transformed

The **one-sided Laplace transform** is defined by:

$$X(s) = \mathcal{L}[x(t)] = \int_0^\infty x(t)\, e^{-st}\, dt \qquad \text{(Eq. 2.3-1)}$$

The **inverse Laplace transform** is defined by:

$$x(t) = \mathcal{L}^{-1}[X(s)] = \frac{1}{2\pi j} \int_C X(s)\, e^{st}\, ds \qquad \text{(Eq. 2.3-2)}$$

where C is a contour of integration chosen so as to include all singularities of X(s).

Fig. 2-9. One-sided Laplace transform.

using this definition, an algebraic equation in the complex frequency domain will result. This equation can then be solved for the desired quantity and transformed back into the time domain by using the inverse Laplace transform. Direct evaluation of the Laplace inversion integral involves the calculus of complex variables (yech!), but fortunately, in the analysis of linear systems, the calculus can be completely avoided by using a number of well-known transform pairs and transform properties presented in Tables 2-1 and 2-2 respectively.

Example 2-1. Using the transform pairs and properties in Tables 2-1 and 2-2, find the Laplace transform of the pulse function shown in Fig. 2-10.

Solution. The given pulse can be represented as a sum of two shifted step functions as in Equation 2.3-22. The additive property given by Equation 2.3-15 indicates that the transform of this sum will be the sum of the transforms due to each of the individual terms. Application of Equations 2.3-4 and 2.3-19 produces the final result shown in 2.3-24.

2.4 SYSTEM ANALYSIS VIA TRANSFORM TECHNIQUES

The Laplace transform has spawned a number of very useful system analysis tools such as transfer functions, pole-zero plots,

Table 2-1. Some Common Laplace Transform Pairs.

x(t)	X(s)	
1	$\frac{1}{s}$	(Eq. 2.3-3)
$u_1(t)$	$\frac{1}{s}$	(Eq. 2.3-4)
$\delta(t)$	1	(Eq. 2.3-5)
t	$\frac{1}{s^2}$	(Eq. 2.3-6)
t^n	$\frac{n!}{s^{n+1}}$	(Eq. 2.3-7)
$\sin \omega t$	$\frac{\omega}{s^2 + \omega^2}$	(Eq. 2.3-8)
$\cos \omega t$	$\frac{s}{s^2 + \omega^2}$	(Eq. 2.3-9)
e^{-at}	$\frac{1}{s+a}$	(Eq. 2.3-10)
$e^{-at} \sin \omega t$	$\frac{\omega}{(s+a)^2 + \omega^2}$	(Eq. 2.3-11)
$e^{-at} \cos \omega t$	$\frac{s+a}{(s+a)^2 + \omega^2}$	(Eq. 2.3-12)

and frequency response characteristics. These tools are each described in the following sections, and the relationships between
them are summarized in Fig. 2-11.

2.4.1 Transfer Functions. The transfer function of a system relates the spectrum of the input signal to the spectrum of the
corresponding output signal. Specifically, the transfer function is
defined as the Laplace transform of the output divided by the
Laplace transform of the input as in Equation 2.4-1. As shown in
Equation 2.4-2, the transfer function is also equal to the Laplace
transform of the system's impulse response. If the transfer function H(s) of a system is known (either by design or from previous
testing), the system's output response y(t) to a particular input x(t)

Table 2-2. Properties of the Laplace Transform.

Time function	Transform	
$x(t)$	$X(s)$	(Eq. 2.3-13)
$a\,x(t)$	$a\,X(s)$	(Eq. 2.3-14)
$y(t) + x(t)$	$Y(s) + X(s)$	(Eq. 2.3-15)
$\dfrac{d}{dt}\,x(t)$	$sX(s) - x(0+)$	(Eq. 2.3-16)
$\displaystyle\int_0^t x(\tau)\,d\tau$	$\dfrac{X(s)}{s}$	(Eq. 2.3-17)
$e^{-at}\,x(t)$	$X(s+a)$	(Eq. 2.3-18)
$u_1(t-\tau)\,x(t-\tau)$	$e^{-\tau s}\,X(s)$	(Eq. 2.3-19)
$\displaystyle\int_0^t h(t-\tau)\,x(\tau)\,d\tau$	$H(s)\,X(s)$	(Eq. 2.3-20)
$x\!\left(\dfrac{t}{a}\right)$	$a\,X(as)$	(Eq. 2.3-21)

can be determined by multiplying H(s) by the Laplace transform of the input and then taking the inverse transform of the resulting product as in Equation 2.4-3. Transfer functions which are expressible as a ratio of polynomials can be inverse transformed into time functions using the Heaviside expansion discussed later.

2.4.2 Poles, Zeros, Damping, and Stability. A Laplace transfer function for a realizable network can always be expressed as a ratio of polynomials as shown in Equation 2.4-4. The roots of the numerator are called *zeros*, while roots of the denominator are called *poles*. If the numerator has a repeated root which appears n times, the corresponding zero is called an *n-th order zero*. Like-

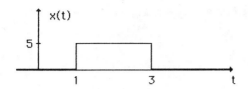

$$x(t) = 5(u_1(t-1) - u_1(t-3)) \qquad (Eq.2.3-22)$$

$$X(s) = 5 [\mathcal{L}[u_1(t-1)] + (-5 [\mathcal{L}[u_1(t-3)]) \qquad (Eq.2.3-23)$$

$$= \frac{5e^{-s}}{s} - \frac{5e^{-3s}}{s} \qquad (Eq.2.3-24)$$

Fig. 2-10. Example of Laplace transformation using tables.

wise, a root appearing n times in the denominator will correspond to an *n-th order pole*. Poles and zeros corresponding to single non-repeated roots are called *simple* poles or zeros as appropriate. Pole and zero locations are often plotted on a graph of the s-plane as shown in Fig. 2-13. In systems which are real functions of time, complex poles of the transfer function will always occur in complex conjugate pairs. Real-valued poles can occur singly or in pairs.

Important insights into a system's behavior can be gained from the pole and zero locations. A pole located at $s = \sigma + j\omega$ corresponds to a natural response component of the form $Ke^{st} = Ke^{(\sigma + j\omega)}$. This complex exponential can take on several forms depending on the values of s as shown in Table 2-3. As shown, systems having poles in the right half of the s-plane are unstable. A system with poles on the $j\omega$ axis is marginally stable if they are first-order poles or unstable if they are multiple-order poles. Zeroes can usually lie anywhere on the s-plane without affecting stability. However their locations play an important role in the steady-state frequency response of a system since input components occurring at a frequency corresponding to a zero will be greatly attenuated by the system.

2.4.3 Magnitude Response. The magnitude response of a CTLTI system is simply the magnitude of the system's steady-state response $H(j\omega)$. This can be obtained in several ways. If $H(j\omega)$ is already in the magnitude and phase form as given by Equation 2.4-5 (Fig. 2-14), then $|H(j\omega)|$ can be found by inspection. However,

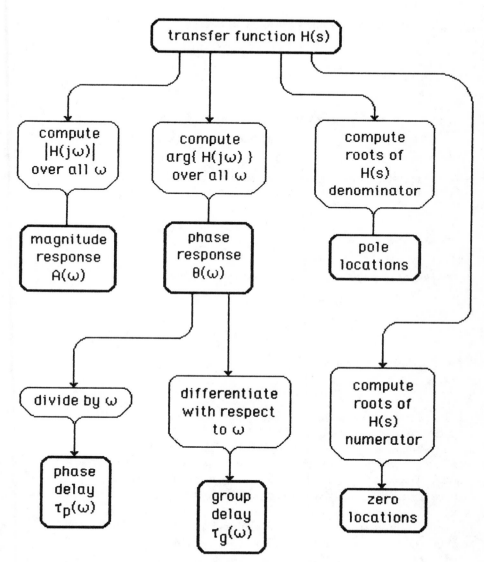

Fig. 2-11. Relationships between various system analysis tools.

if H(jω) is in the rectangular form given by Equation 2.4-6, then |H(jω)| is obtained by using Equation 2.4-7. If the system transfer function is available in factored form as in Equation 2.4-8, then |H(jω)| can be obtained by replacing each factor with its absolute value when evaluated at s = jω as in Equation 2.4-9.

A system's transfer function, H(s) is given by

$$H(s) = \frac{Y(s)}{X(s)} \qquad \text{(Eq. 2.4-1)}$$

where
   $Y(s)$ is the Laplace transform of the output $y(t)$

   $X(s)$ is the Laplace transform of the input $x(t)$.

The transfer function H(s) can be obtained by taking the Laplace transform of the impulse response $h(t)$:

$$H(s) = \mathcal{L}[h(t)] \qquad \text{(Eq. 2.4-2)}$$

The response $y(t)$ due to a particular input $x(t)$ can be computed as:

$$y(t) = \mathcal{L}^{-1}\left[H(s)\ \mathcal{L}[x(t)]\right] \qquad \text{(Eq. 2.4-3)}$$

Fig. 2-12. Laplace transfer functions.

2.4.4 Phase Response. Consider once again a system having a steady-state response given by Equation 2.4-5. This system's phase response is defined as $\theta(\omega)$. As with the magnitude response, there are several ways in which the phase response can be computed. If $H(j\omega)$ is in the form of Equation 2.4-5, $\theta(\omega)$ can be obtained by inspection. If $H(j\omega)$ is given in rectangular form then $\theta(\omega)$ is obtained by using Equation 2.4-10. If the system transfer function $H(s)$ is given in factored form as in Equation 2.4-8, then $\theta(\omega)$ can be obtained by using Equation 2.4-11 with o_i, b_i, c_i, and d_i as indicated in Fig. 2-15.

2.4.5 Phase Delay. In many applications it is often necessary to know how much delay will be experienced by a sinusoid of a particular frequency when passing through a system. In general, each frequency will experience a different delay. Collectively these delays are called *phase delay* or *carrier delay* and are given by Equation 2.4-12. (Assuming that negative values of $\theta(\omega)$

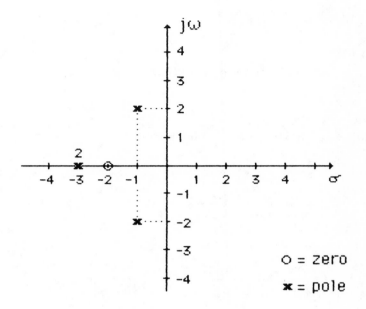

$$H(s) = \frac{s + 2}{s^4 + 4s^3 + 2s^2 + 12s + 45}$$

(Eq. 2.4-4)

$$= \frac{s+2}{(s+1+2j)(s+1-2j)(s+3)(s+3)}$$

zero: $s = -2$

poles: $s = -1+2j$, $s = -1-2j$, $s = -3$, $s = -3$

Fig. 2-13. Plot of pole and zero locations.

$$H(j\omega) = |H(j\omega)| \, e^{j\theta(\omega)} \qquad\qquad \text{(Eq. 2.4-5)}$$

$$= \text{Re}\{ H(j\omega) \} + j \, \text{Im}\{ H(j\omega) \} \qquad\qquad \text{(Eq. 2.4-6)}$$

$$|H(j\omega)| = \sqrt{ [\, \text{Re}\{ H(j\omega) \} \,]^2 + [\, \text{Im}\{ H(j\omega) \} \,]^2 } \qquad \text{(Eq. 2.4-7)}$$

$$H(s) = H_0 \, \frac{(s-z_1)(s-z_2)(s-z_3)\dots(s-z_m)}{(s-p_1)(s-p_2)(s-p_3)\dots(s-p_n)} \qquad \text{(Eq. 2.4-8)}$$

$$|H(j\omega)| = H_0 \, \frac{|j\omega-z_1|\cdot|j\omega-z_2|\cdot|j\omega-z_3|\dots|j\omega-z_m|}{|j\omega-p_1|\cdot|j\omega-p_2|\cdot|j\omega-p_3|\dots|j\omega-p_n|}$$

$$\text{(Eq. 2.4-9)}$$

Fig. 2-14. Formulas for computing a system's magnitude response.

are interpreted as a phase lag of the output relative to the input, the minus sign in Equation 2.4-12 is needed to yield a positive value of delay. Some sources may define τ_p under the opposite sign convention.) As shown in Fig. 2-16, the phase delay at a frequency ω_i is equal to the negative slope of a secant from the phase plot origin to the phase response at ω_i.

2.4.6 Group Delay. The spectrum of a signal such as an amplitude-modulated sinusoidal carrier contains components or *sidelobes* due to the modulation. The interaction of the different phase delays for these sidelobes and for the carrier will cause the modulation envelope to be delayed by an amount which is in general not equal to the phase delays of the spectral components. The delay induced in the envelope is called *envelope delay* or *group delay* and is given by Equation 2.4-13. As shown in Fig. 2-17, the group delay at a frequency ω_i is equal to the negative slope of a tangent to the phase response at ω_i. If the bandwidth of a modulated signal is contained entirely within a frequency region exhibiting constant group delay, the envelope will be delayed by an amount equal to

τ_g while the carrier of frequency ω_c will be delayed by an amount equal to the phase delay $\tau_p(\omega_c)$. If the group delay is not constant over the entire bandwidth of the signal, distortion of the envelope will occur.

THE HEAVISIDE EXPANSION

The Heaviside expansion provides a direct computational method for performing an inverse Laplace transform on a complex-frequency function which is expressed as a ratio of polynominals in s, where the order of the denominator polynomial exceeds the order of the numerator polynomial. The general case expansion is

Table 2-3. Impact of Various Types of Pole Values.

Pole Type	Response Component	Description
single real, negative	decaying exponential	stable
single real, positive	divergent exponential	divergent instability
real pair, negative, unequal	decaying exponential	overdamped (stable)
real pair, negative, equal	decaying exponential	critically damped (stable)
real pair, positive	divergent exponential	divergent instability
complex conjugate pair with negative real parts	exponentially decaying sinusoid	underdamped (stable)
complex conjugate pair with zero real parts	sinusoid	undamped (marginally stable)
complex conjugate pair with positive real parts	exponentially saturating sinusoid	oscillatory instability

$$\theta(\omega) = \tan^{-1} \frac{\text{Im}\{ H(j\omega) \}}{\text{Re}\{ H(j\omega) \}} \qquad \text{(Eq. 2.4-10)}$$

$$\theta(\omega) = \tan^{-1} \frac{d_1 - \omega}{c_1} + \tan^{-1} \frac{d_2 - \omega}{c_2} \ldots + \tan^{-1} \frac{d_m - \omega}{c_m}$$

$$- \tan^{-1} \frac{b_1 - \omega}{a_1} - \tan^{-1} \frac{b_2 - \omega}{a_2} \ldots - \tan^{-1} \frac{b_n - \omega}{a_n}$$

$$\text{(Eq. 2.4-11)}$$

where

$$a_i = \text{Re}\{ p_i \} \qquad c_i = \text{Re}\{ z_i \}$$
$$b_i = \text{Im}\{ p_i \} \qquad d_i = \text{Im}\{ z_i \}$$
$$p_i = a_i + j b_i \qquad z_i = c_i + j b_i$$

Fig. 2-15. Formulas for computing a system's phase response.

$$\tau_p(\omega) = \frac{-\theta(\omega)}{\omega} \qquad \text{(Eq. 2.4-12)}$$

where $\theta(\omega)$ is the phase response of the system
transfer function

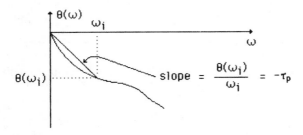

Fig. 2-16. Phase delay.

$$\tau_g(\omega) = \frac{-d}{dt} \theta(\omega) \qquad\qquad\text{(Eq. 2.4-13)}$$

where $\theta(\omega)$ is the phase response of the system transfer function

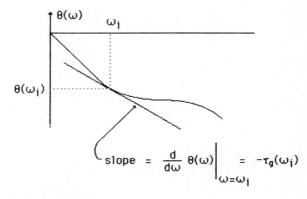

$$\text{slope} = \frac{d}{d\omega} \theta(\omega)\bigg|_{\omega=\omega_i} = -\tau_g(\omega_i)$$

Fig. 2-17. Group delay.

if

$$H(s) = K_0 \frac{P(s)}{Q(s)}$$

where

$$Q(s) = (s-s_1)^{m_1} (s-s_2)^{m_2} \dots (s-s_n)^{m_n}$$

then

$$\mathcal{L}^{-1}\big[H(s)\big] = K_0 \sum_{r=1}^{n} \sum_{k=1}^{m_r} \big[K_{rk} t^{m_r-k} e^{s_r t} \big]$$

where

$$K_{rk} = \frac{1}{(k-1)! \, (m_r - k)!} \frac{d^{k-1}}{ds^{k-1}} \left[\frac{(s-s_r)m_r P(s)}{Q(s)} \right]_{s=s_r}$$

Fig. 2-18. General Heaviside expansion.

if

$$C(s) = \frac{A(s)}{B(s)} \quad ; B(s) \neq 0$$

then

$$\frac{d}{ds} C(s) = \frac{B(s) \frac{d}{ds} A(s) - A(s) \frac{d}{ds} B(s)}{[B(s)]^2}$$

Fig. 2-19. Formula for taking derivatives used in calculation of the Heaviside expansion.

given in Fig. 2-18. The theorem given in Fig. 2-19 is useful for taking the derivatives required to calculate K_{rk}. In cases involving only simple poles, the Heaviside expansion can be simplified to the form shown in Fig. 2-20.

if

$$H(s) = K_0 \frac{P(s)}{Q(s)}$$

where

$$Q(s) = (s-s_1)(s-s_2)(s-s_3) \ldots (s-s_n) \, ;$$

$$s_1 \neq s_2 \neq s_3 \neq \ldots \neq s_n$$

then

$$L^{-1}[H(s)] = K_0 \sum_{r=1}^{n} K_r e^{s_r t}$$

where

$$K_r = \left[\frac{(s-s_r) P(s)}{Q(s)} \right]_{s=s_r}$$

Fig. 2-20. Simplified Heaviside expansion for the case of simple poles.

Chapter 3

Filter Fundamentals

I N MANY SIGNAL PROCESSING APPLICATIONS, THE DESIRED signal and the undesired noise or contaminating signal will be concentrated in different parts of the frequency spectrum. By using filters that allow the desired frequencies to pass while attenuating the undesired frequencies, much of the noise can be eliminated. This chapter will examine the mathematical model of an ideal filter and various physically realizable approximations to this nonrealizable ideal. All of the filters that we will study are linear systems, and as such they are subject to all of the analytical tools presented in Chapter 2.

3.1 IDEAL FILTERS

One possible approach for filter design involves the rectangular magnitude responses shown in Fig. 3-1. The desired frequencies are passed with no attenuation, while the undesired frequencies are completely blocked. Such filters are called *ideal filters* and if they could be implemented, they would enjoy widespread use. Unfortunately, analysis of the impulse and step responses of ideal filters indicate that they are noncausal and therefore not realiza-

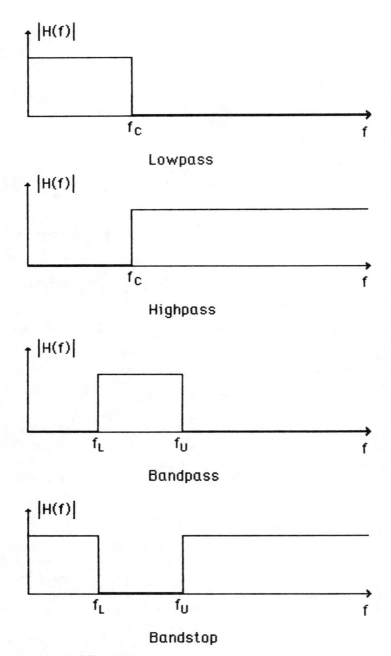

Fig. 3-1. Ideal filter responses.

ble. However, there are practical filter designs that approximate the ideal filter characteristics and are realizable. Each of the major types—Butterworth, Chebyshev, Bessel, and elliptical—optimizes a different aspect of the approximation.

Equations describing the various characteristics of an ideal lowpass filter are given in Fig. 3-2. Since the phase delay (3.1-4) and group delay (3.1-5) are constant and equal, any steady-state input whose spectrum lies completely within the passband will be delayed by $n\pi/2\omega_c$ and amplified by k when passing through the filter, while spectral components outside the passband will be completely stopped by the filter. The impulse response plotted in Fig. 3-3 is obtained by taking the inverse Laplace transform of (3.1-1) to yield (3.1-6). Since the impulse response begins before the application of the impulse input, ideal lowpass filters are obviously noncausal and therefore not physically realizable.

3.2 LOWPASS FILTER RESPONSE DATA

In the remainder of this chapter we will encounter a number of different approximations to the ideal filter response. To make comparisons between the various filter types more convenient we first need to define a few terms and establish some rules for *normalizing* the filter response data.

3.2.1 Magnitude Response Features of Lowpass Filters. The magnitude response of a practical lowpass filter will usually have one of the four general shapes shown in Figs. 3-4 through 3-7. In all four cases the filter characteristics divide the frequency spectrum into three general regions as shown. The *passband* extends from dc up to the cutoff frequency ω_c. The *transition band* extends from ω_c up to the beginning of the stopband at ω_1, and of course the *stopband* extends upward from ω_1 to infinity. The *cutoff frequency* ω_c is usually defined as the frequency at which the amplitude response falls to a value 3 dB below the peak passband value. For a peak passband value of A_o, an attenuation of 3 dB (or more precisely 3.0103 dB) will produce an amplitude of $A_o/\sqrt{2}$. This corresponds to the half-power point, since power is proportional to A_o^2. Defining the frequency ω_1 which marks the beginning of the stopband is not quite so straightforward. In Fig. 3-4 or 3-5 there really isn't any particular feature that indicates just where ω_1 should be located. For some applications the specification can be quite arbitrary—such as a requirement that there be some specified minimum amount of attenuation at some particular

Transfer function:

$$H(s) = \begin{cases} K\,e^{-st} & \text{for s within the passband} \\ 0 & \text{for s outside the passband} \end{cases}$$

(Eq. 3.1-1)

Magnitude response:

$$|H(j\omega)| = \begin{cases} K & \text{for } \omega \text{ within the passband} \\ 0 & \text{for } \omega \text{ outside the passband} \end{cases}$$

(Eq. 3.1-2)

Phase response:

$$\theta(\omega) = \frac{-n\pi\omega}{2\omega_c}$$

(Eq. 3.1-3)

Phase delay:

$$\tau_p(\omega) = \frac{-\theta(\omega)}{\omega} = \frac{n\pi}{2\omega_c}$$

(Eq. 3.1-4)

Group delay:

$$\tau_g(\omega) = \frac{d}{d\omega}\theta(\omega) = \frac{n\pi}{2\omega_c}$$

(Eq. 3.1-5)

Fig. 3-2. Characteristics of an ideal lowpass filter.

frequency. Should this frequency be defined as ω_1? The answer is really not very clear. The commonly adopted solution involves specifying a (possibly arbitrary) *minimum stopband loss* α_2 (or conversely a maximum stopband amplitude A_2) and then defining ω_1 as the first frequency at which the loss exceeds and subsequently

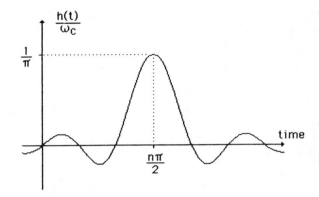

$$h(t) = \frac{\omega_c}{\pi} \frac{\sin(\omega_c t - n\pi/2)}{\omega_c t - n\pi/2} \qquad (Eq.\,3.1-6)$$

Fig. 3-3. Impulse response of an ideal lowpass filter.

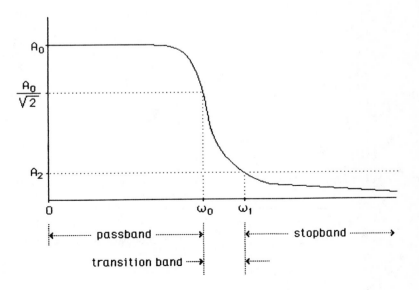

Fig. 3-4. Monotonic magnitude response of a practical lowpass filter.

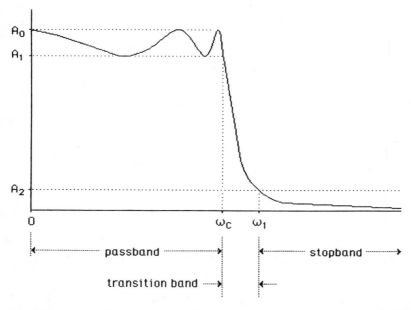

Fig. 3-5. Magnitude response of a practical lowpass filter with ripples in the passband.

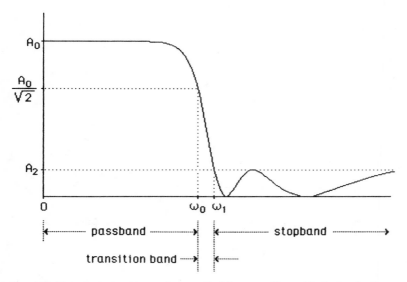

Fig. 3-6. Magnitude response of a practical lowpass filter with ripples in the stopband.

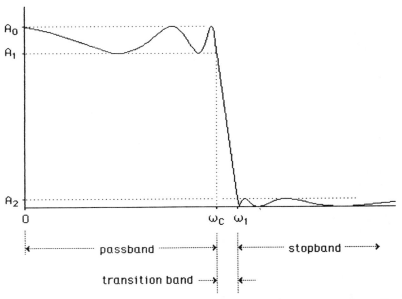

Fig. 3-7. Magnitude response of a practical lowpass filter with ripples in the passband and stopband.

continues to exceed α_2. The width W_T of the transition band is equal to $\omega_c - \omega_1$. The quantity W_T / ω_c is sometimes called the *normalized transition width*. In the case of response shapes like those shown in Figs. 3-6 and 3-7, the minimum stopband loss is clearly defined by the peaks of the stopband ripples.

The decreasing monotonic response of Fig. 3-4 is characteristic of Butterworth lowpass filters which are discussed in Chapter 4. The response having ripples in the passband (Fig. 3-5) is characteristic of Chebyshev filters and the response having ripples in the stopband (Fig. 3-6) is characteristic of inverse Chebyshev filters. Chebyshev filters are discussed in Chapter 5. The response shown in Fig. 3-7 which has ripples in both the passband and stopband is characteristic of elliptical filters which are somewhat difficult to analyze and which will not be covered further in this book. The interested reader should consult Chapter 5 of *Johnson, et al, 1980*, which contains a good discussion of elliptical filters.

3.2.2 Frequency Scaling of Lowpass Magnitude Responses. In plots of practical filter responses, the frequency axes are almost universally plotted on logarithmic scales. Magnitude response curves for lowpass filters are scaled so that the cutoff frequency ω_c occurs at a convenient frequency such as 1 radian per second, 1 Hz, or 1 kHz. A single set of such normalized response curves can then be denormalized to fit any particular cutoff requirement. Often it is computationally easier to normalize the design requirements for comparison to the normalized response curves, rather than to denormalize the curves for comparison to the original design requirements.

3.2.3 Magnitude Scaling. The vertical axis of a filter's magnitude response can be presented in several different forms. In most theoretical development, the usual presentation involves plotting the amplitude response on a linear scale, while in practical design situations it is often more convenient to work with plots of amplitude response or attenuation in dB using a high resolution linear scale in the passband and a lower resolution linear scale in the stopband. This allows details of the basspand response to be shown as well as large attenuation values deep into the stopband. In all cases, the vertical data is usually normalized to present a 0 dB amplitude or 0 dB attenuation at the peak of the passband.

3.2.4 Phase Response. The phase response (not to be confused with phase delay) is usually plotted as a phase angle in degrees or radians versus frequency. By adding or subtracting the appropriate number of full-cycle offsets (i.e., 2π radians or 360°), the phase response can be presented either as a single curve extending over several full cycles (Fig. 3-8) or as an equivalent set of curves each of which extends over a single cycle (Fig. 3-9). The format of Fig. 3-9 allows a more compact plot of the phase response over a large range of frequencies.

3.2.5 Step Response. Normalized step response plots are obtained by computing the step response from the normalized transfer function. The inherent scaling of the time axis will thus depend upon the transient characteristics of the normalized filter. The amplitude axis scaling is not dependent upon normalization. The usual lowpass presentation will require that the response be denormalized by dividing the frequency axis by some form of the cutoff frequency.

3.2.6 Impulse Response. Normalized impulse response plots are obtained by computing the impulse response from the normalized transfer function. Since an impulse response will always have an area of unity, both the time axis and the amplitude axis will exhibit inherent scaling which depends upon the transient characteristics of the normalized filter. The usual lowpass presentation will require that the response be denormalized by multiply-

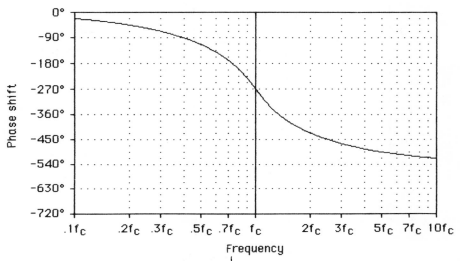

Fig. 3-8. Phase response plot extending over multiple cycles.

ing the amplitude by some form of the cutoff frequency and dividing the time axis by the same factor.

3.3 HIGHPASS FILTERS

So far we have discussed lowpass filters and some of their characteristics, but what about the other types—highpass, bandpass, and bandstop? Rather than design completely new filters, it is much easier to take existing lowpass designs and transform them into corresponding highpass, bandpass, or bandstop configurations.

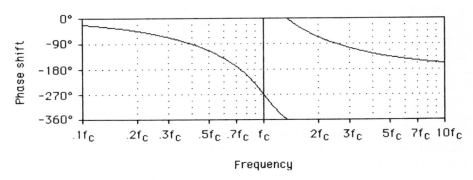

Fig. 3-9. Phase response plot confined to a single-cycle range.

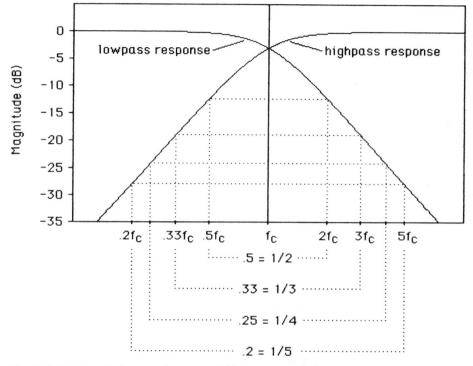

Fig. 3-10. Relationship between lowpass and highpass magnitude responses.

Normalized lowpass transfer functions can be converted into corresponding highpass transfer functions by simply replacing each occurrence of "s" with "1/s." This will cause the magnitude response plot to be "flipped" around a line at f_c as shown in Fig. 3-10. (Note that this "flip" only works when the frequency is plotted on a logarithmic scale.) Rather than actually trying to draw a flipped response curve, it is much simpler to take the reciprocals of all the important frequencies for the highpass filter in question and then read the appropriate response directly from the lowpass curves. In general, the transient response characteristics of the lowpass filter are *not* preserved by this transformation. Other transformations that preserve transient response at the expense of frequency response do exist. The interested reader should consult page 165 of *Blinchikoff and Zverev, 1976.*

3.4 BANDPASS FILTERS

Bandpass filters are classified as wideband, or narrowband based on the relative width of their passbands. Different methods are used for obtaining the transfer function for each type.

3.4.1 Wideband Bandpass Filters. Wideband bandpass filters can be realized by cascading a lowpass filter and a highpass filter. This approach will be acceptable as long as the bandpass bandwidth is relatively wide and the magnitude responses of the filters used exhibit relatively sharp transitions from the passband to cutoff. Relatively narrow bandwidths and/or gradual rolloffs which begin well within the passband, can cause a significant centerband loss as shown in Fig. 3-11. In situations where such losses are unacceptable, other bandpass filter realizations must be used. A general rule-of-thumb is to use narrowband techniques for passbands which are an octave or smaller ($(f_H/f_L) \leq 2$).

3.4.2 Narrowband Bandpass Filters. A normalized lowpass filter can be converted into a normalized narrowband bandpass filter by substituting "$(s - (1/s))$" for "s" in the lowpass transfer function. The center frequency of the resulting bandpass filter will be at the cutoff frequency of the original lowpass filter and the passband will be symmetric about the center frequency when plotted on a logarithmic frequency scale. At any particular attenuation level, the bandwidth of the bandpass filter will equal the frequency at which the lowpass filter exhibits the same attenuation. (See Fig.

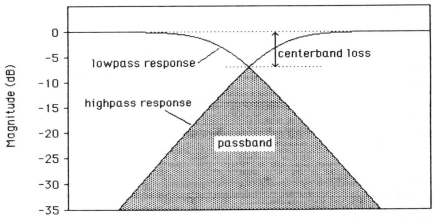

Fig. 3-11. Centerband loss in a bandpass filter realized by cascading lowpass and highpass filters.

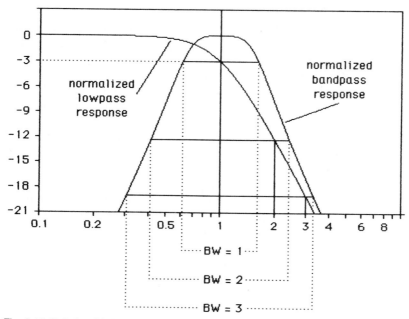

Fig. 3-12. Relationship between lowpass and bandpass magnitude responses.

3-12.) This particular bandpass transformation preserves the magnitude response shape of the lowpass prototype, but distorts the transient responses.

3.5 BANDSTOP FILTERS

A normalized lowpass filter can be converted into a normalized bandstop filter by substituting "$s/(s^2-1)$" for "s" in the lowpass transfer function. The center frequency of the resulting bandstop filter will be at the cutoff frequency of the original lowpass filter and the stopband will be symmetrical about the center frequency when plotted on a logarithmic frequency scale. At any particular attenuation level, the width of the stopband will be equal to the reciprocal of the frequency at which the lowpass filter exhibits the same attenuation (see Fig. 3-13).

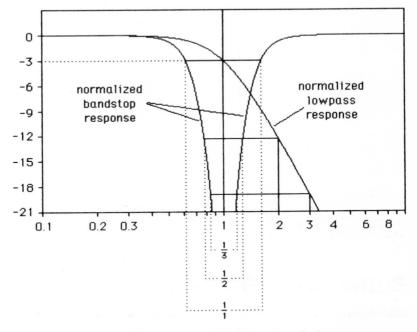

Fig. 3-13. Relationship between lowpass and bandstop magnitude responses.

Chapter 4

Butterworth Filters

B UTTERWORTH LOWPASS FILTERS ARE DESIGNED TO HAVE an amplitude response characteristic that is as flat as possible at low frequencies and monotonically decreasing.

4.1 TRANSFER FUNCTION

The general expression for the transfer function of an n-th order Butterworth lowpass filter is given by Equation 4.1-1 of Fig. 4-1. To generate the transfer function of a specific order from this expression, just pick the desired order n and plug it into Equation 4.1-2 to determine s_1, s_2, s_3, . . . s_n. Then form the denominator product as indicated.

Example 4-1. Determine the transfer function for a lowpass third-order Butterworth filter.

Solution. The third-order transfer function will have the form shown in Fig. 4-2. The values shown for s_1, s_2, s_3 are obtained from Equation 4.1-2(b).

From the form of (4.1-1) we can see that an n-th order Butterworth filter will always have n poles and no finite zeros. Also true, but not quite so obvious, is the fact that these poles lie at equally spaced points on the left half of a circle in the s plane. Pole values,

$$H(s) = \frac{1}{\displaystyle\prod_{i=1}^{n} (s-s_i)}$$

$$= \frac{1}{(s-s_1)(s-s_2)\ldots.(s-s_n)} \qquad \text{(Eq. 4.1-1)}$$

$$s_i = e^{j\pi[(2i+n-1)/2n]} \qquad \text{(Eq. 4.1-2a)}$$

$$= \cos\left[\pi\,\frac{2i+n-1}{2n}\right] + j\sin\left[\pi\,\frac{2i+n-1}{2n}\right]$$

$$\text{(Eq. 4.1-2b)}$$

$$H(s) = \prod_{i=1}^{n/2} \frac{k_i}{s^2 + b_i s + 1} \qquad \text{(Eq. 4.1-3)}$$

$$H(s) = \frac{k_0}{s+1} \prod_{i=1}^{(n-1)/2} \frac{k_i}{s^2 + b_i s + 1} \qquad \text{(Eq. 4.1-4)}$$

Fig. 4-1. Transfer function for lowpass Butterworth filters.

$$H(s) = \frac{1}{(s-s_1)(s-s_2)(s-s_3)}$$

$$s_1 = \cos(2\pi/3) + j \sin(2\pi/3) = -0.5 + 0.866j$$

$$s_2 = e^{j\pi} = \cos(\pi) + j \sin(\pi) = -1$$

$$s_3 = \cos(4\pi/3) + j \sin(4\pi/3) = -0.5 - 0.866j$$

$$H(s) = \frac{1}{(s + .5 - .866j)(s + 1)(s + .5 - .866j)}$$

$$= \frac{1}{s^3 + 2s^2 + 2s + 1}$$

Fig. 4-2. Equations for Example 4-1.

transfer functions, and s-plane plots for Butterworth filters of orders 2 through 8 are shown in Figs. 4-3 through 4-9. Higher-order filters are often implemented as cascaded first- and second-order sections. When designing these sections, it is very convenient to have the transfer function in the form of Equation 4.1-3 in which each pair of complex conjugate poles is combined into a single second-order filter section. If the overall filter order is odd, the one real pole will stand alone as a first-order section and Equation 4.1-4 will apply. Table 4-1 lists the appropriate section coefficients for Butterworth filters of orders 2 through 8. The design procedures of Chapter 6 require the filter coefficients in this form.

4.2 FREQUENCY RESPONSE

Figures 4-10 through 4-12 show respectively, the passband amplitude response, stopband amplitude response, and phase response for Butterworth filters of orders 2 through 8. (The BASIC programs used to generate these various response plots are presented at the end of this chapter. These frequency response plots are normal-

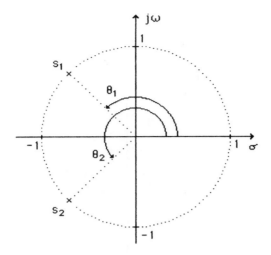

$$H(s) = \frac{K}{(s-s_1)(s-s_2)}$$

$s_1 = -0.707107 + 0.707107\,j$ $\theta_1 = 135.0° = 2.35619$ rad

$s_2 = -0.707107 - 0.707107\,j$ $\theta_2 = 225.0° = 3.92699$ rad

Fig. 4-3. Pole locations for a second-order Butterworth LPF.

ized for a cutoff frequency of 1 Hz. To denormalize them, simply multiply the frequency axis by the desired cutoff frequency f_c.

Example 4-2. Use Figs. 4-11 and 4-12 to determine the amplitude and phase response at 800 Hz of a sixth-order Butterworth lowpass filter having a cutoff frequency of 400 Hz.

Solution. If you set $f_c = 400$ we can denormalize the $n = 6$ response of Fig. 4-11 to obtain the response shown in Fig. 4-13. From this you can easily determine that the amplitude response at 800 Hz is approximately − 36 dB. Likewise, you can denormalize the $n = 6$ response of Fig. 4-12 to obtain the response shown in Fig. 4-14. From this you can determine that the phase response at 800 Hz is approximately − 425°.

4.3 DETERMINATION OF MINIMUM FILTER ORDER

Usually in the real world the order of the desired filter is not

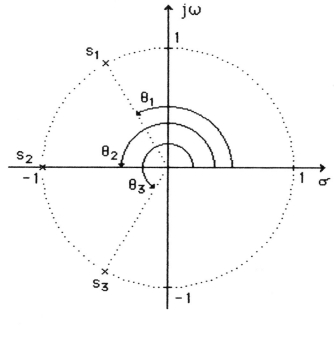

$$H(s) = \frac{K}{(s-s_1)(s-s_2)(s-s_3)}$$

$s_1 = -0.5 + 0.866025\ j$ $\theta_1 = 120.0° = 2.09440$ rad

$s_2 = -1$ $\theta_2 = 180.0° = 3.14159$ rad

$s_3 = -0.5 - 0.866025\ j$ $\theta_3 = 240.0° = 4.18879$ rad

Fig. 4-4. Pole locations for a third-order Butterworth LPF.

given as in Example 4-2, but instead must be based on the required performance of the filter. For lowpass Butterworth filters, the minimum order n which will insure a magnitude of A_2 or lower at all frequencies ω_1 and above can be obtained by using Equation 4.3-1.

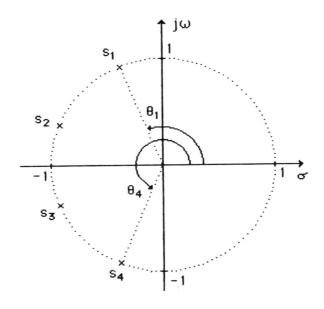

$$H(s) = \frac{K}{(s-s_1)(s-s_2)(s-s_3)(s-s_4)}$$

$s_1 = -0.382683 + 0.923880\ j$ $\theta_1 = 112.5° = 1.96350$ rad

$s_2 = -0.923880 + 0.382683\ j$ $\theta_2 = 157.5° = 2.74889$ rad

$s_3 = -0.923880 - 0.382683\ j$ $\theta_3 = 202.5° = 3.53429$ rad

$s_4 = -0.382683 - 0.923880\ j$ $\theta_4 = 247.5° = 4.31969$ rad

Fig. 4-5. Pole locations for a fourth-order Butterworth LPF.

(Note that the value of A_2 will be negative, thus canceling the minus sign in the numerator exponent.)

4.4 IMPULSE RESPONSE

Impulse responses for lowpass Butterworth filters are shown in Figs. 4-16 and 4-17. (The computer programs used to generate these responses are presented and discussed at the end of this chapter.) These responses are normalized for lowpass filters having a cutoff frequency equal to 1 radian per second. To denormalize the response, divide the time axis by the desired cutoff frequency ω_c = $2\pi f_c$ and multiply the amplitude axis by the same factor.

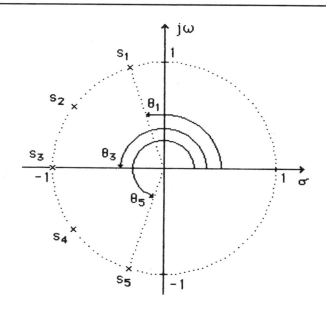

$$H(s) = \frac{K}{(s-s_1)(s-s_2)(s-s_3)(s-s_4)(s-s_5)}$$

$s_1 = -0.309017 + 0.951057j$ $\qquad$ $\theta_1 = 108.0° = 1.88496$ rad

$s_2 = -0.809017 + 0.587785j$ $\qquad$ $\theta_2 = 144.0° = 2.51327$ rad

$s_3 = -1$ $\qquad$ $\theta_3 = 180.0° = 3.14159$ rad

$s_4 = -0.809017 - 0.587785j$ $\qquad$ $\theta_4 = 216.0° = 3.76991$ rad

$s_5 = -0.309017 - 0.951057j$ $\qquad$ $\theta_5 = 252.0° = 4.39823$ rad

Fig. 4-6. Pole locations for a fifth-order Butterworth LPF.

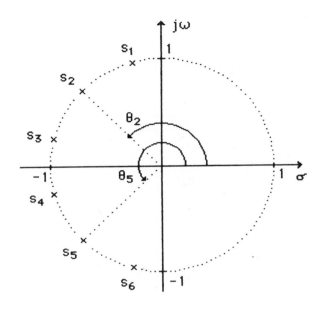

$$H(s) = \frac{K}{(s-s_1)(s-s_2)(s-s_3)(s-s_4)(s-s_5)(s-s_6)}$$

$s_1 = -0.258819 + 0.965926\ j$ $\theta_1 = 105.0° = 1.83260$ rad

$s_2 = -0.707107 + 0.707107\ j$ $\theta_2 = 135.0° = 2.35619$ rad

$s_3 = -0.965926 + 0.258819\ j$ $\theta_3 = 165.0° = 2.87979$ rad

$s_4 = -0.965926 - 0.258819\ j$ $\theta_4 = 195.0° = 3.40339$ rad

$s_5 = -0.707107 - 0.707107\ j$ $\theta_5 = 225.0° = 3.92699$ rad

$s_6 = -0.258819 - 0.965926\ j$ $\theta_6 = 255.0° = 4.45059$ rad

Fig. 4-7. Pole locations for a sixth-order Butterworth LPF.

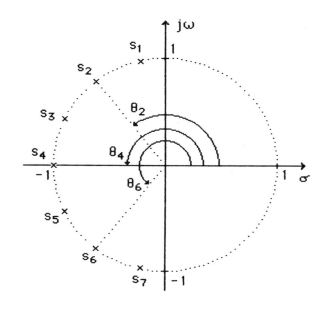

$$H(s) = \frac{K}{(s-s_1)(s-s_2)(s-s_3)(s-s_4)(s-s_5)(s-s_6)(s-s_7)}$$

$s_1 = -0.222521 + 0.974928\,j$ $\qquad$ $\theta_1 = 102.9° = 1.79520$ rad

$s_2 = -0.623490 + 0.781831\,j$ $\qquad$ $\theta_2 = 128.6° = 2.24399$ rad

$s_3 = -0.900969 + 0.433884\,j$ $\qquad$ $\theta_3 = 154.3° = 2.69279$ rad

$s_4 = -1$ $\qquad$ $\theta_4 = 180.0° = 3.14159$ rad

$s_5 = -0.900969 - 0.433884\,j$ $\qquad$ $\theta_5 = 205.7° = 3.59039$ rad

$s_6 = -0.623490 - 0.781831\,j$ $\qquad$ $\theta_6 = 231.4° = 4.03919$ rad

$s_7 = -0.222521 - 0.974928\,j$ $\qquad$ $\theta_7 = 257.1° = 4.48799$ rad

Fig. 4-8. Pole locations for a seventh-order Butterworth LPF.

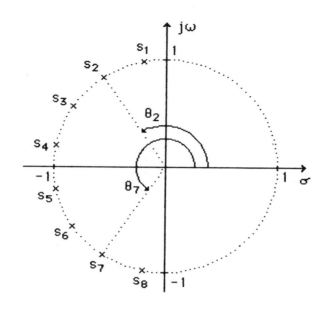

$$H(s) = \frac{K}{(s-s_1)(s-s_2)(s-s_3)(s-s_4)(s-s_5)(s-s_6)(s-s_7)(s-s_8)}$$

$s_1 = -0.195090 + 0.980785\ j$ $\theta_1 = 101.3° = 1.76715$ rad

$s_2 = -0.555570 + 0.831470\ j$ $\theta_2 = 123.8° = 2.15984$ rad

$s_3 = -0.831470 + 0.555570\ j$ $\theta_3 = 146.3° = 2.55254$ rad

$s_4 = -0.980785 + 0.195090\ j$ $\theta_4 = 168.8° = 2.94524$ rad

$s_5 = -0.980785 - 0.195090\ j$ $\theta_5 = 191.3° = 3.33794$ rad

$s_6 = -0.831470 - 0.555570\ j$ $\theta_6 = 213.8° = 3.73064$ rad

$s_7 = -0.555570 - 0.831470\ j$ $\theta_7 = 236.3° = 4.12334$ rad

$s_8 = -0.195090 - 0.980785\ j$ $\theta_8 = 258.8° = 4.51604$ rad

Fig. 4-9. Pole locations for an eighth-order Butterworth LPF.

Order = 2	1.414214
Order = 3	1.000000
Order = 4	0.765367
	1.847759
Order = 5	0.618034
	1.618034
Order = 6	0.517638
	1.414214
	1.931852
Order = 7	0.445042
	1.246980
	1.801938
Order = 8	0.390181
	1.111140
	1.662939
	1.961571

Table 4-1. Coefficients for Butterworth Filters Realized as Cascaded First- and Second-Order Sections.

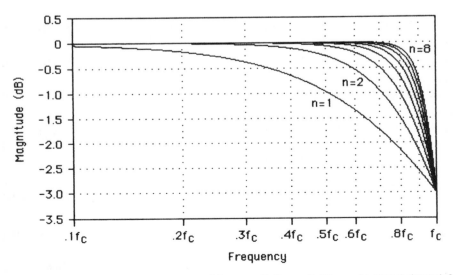

Fig. 4-10. Passband magnitude response of lowpass Butterworth filters of orders 1 through 8.

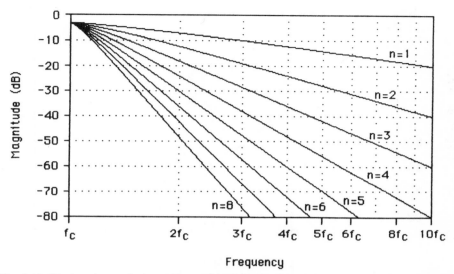

Fig. 4-11. Stopband magnitude response of lowpass Butterworth filters of orders 1 through 8.

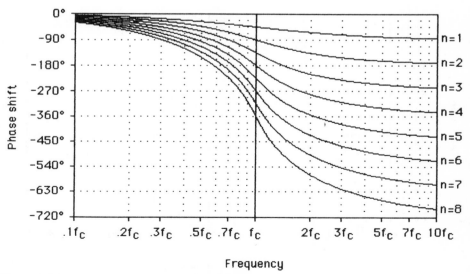

Fig. 4-12. Phase response of lowpass Butterworth filters of orders 1 through 8.

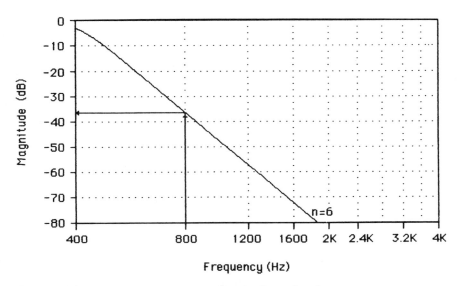

Fig. 4-13. Denormalized magnitude response for Example 4-2.

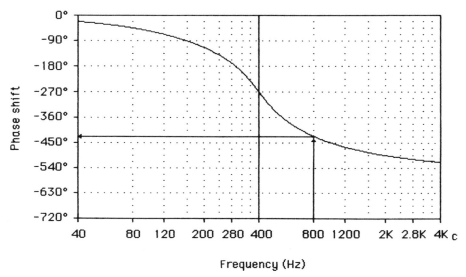

Fig. 4-14. Denormalized phase response for Example 4-2.

74

$$n = \frac{\log (10^{-A_2/10} - 1)}{2 \log (\omega_1/\omega_c)} \qquad \text{(Eq. 4.3-1)}$$

where

ω_c is the 3 dB frequency

ω_1 is the frequency at which the magnitude
response first falls below A_2

Fig. 4-15. Formula for determining the minimum order Butterworth filter needed to obtain a specific response.

Example 4-3. Use Fig. 4-17 to determine the instantaneous amplitude of the output 1.6 msec after a unit impulse is applied to the input of a fifth-order Butterworth LPF having $f_c = 250$ Hz.

Solution. If you set $f_c = 250$ ($\omega_c = 1570.8$) we can denormalize the $n=5$ response of Fig. 4-17 to obtain the response shown in Fig. 4-18. From this you can observe that the response amplitude at $t = 1.6$ msec is approximately 378.

4.5 STEP RESPONSE

Step responses for lowpass Butterworth filters are shown in

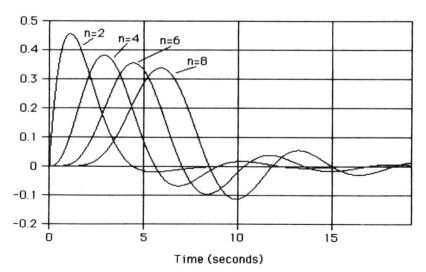

Fig. 4-16. Impulse response of even-order lowpass Butterworth filters.

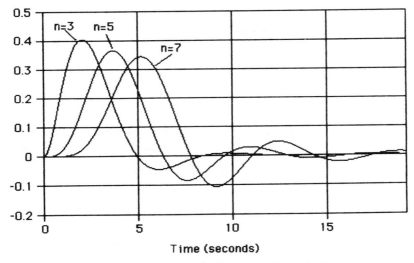

Fig. 4-17. Impulse response of odd-order lowpass Butterworth filters.

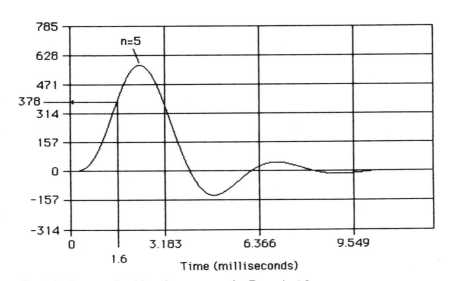

Fig. 4-18. Denormalized impulse response for Example 4-3.

Figs. 4-19 and 4-20. (The computer program used to generate these responses is presented and discussed at the end of the chapter.) These responses are normalized for lowpass filters having a cutoff frequency equal to 1 radians per second. To denormalize the response, divide the time axis by the desired cutoff frequency $\omega_c = 2\pi f_c$.

Example 4-4 Use Fig. 4-20 to determine how long it will take for the step response of a third-order Butterworth LPF ($f_c = 4$ kHz) to first reach 100 percent of its final value.

Solution. If you set $f_c = 4000$ ($\omega_c = 25132.7$), you can denormalize the $n = 3$ response of Fig. 4-20 to obtain the response shown in Fig. 4-21. From this you can observe that the time required for the step response to reach a value of one is approximately 150 microseconds.

COMPUTATION OF BUTTERWORTH TRANSFER FUNCTION DATA

The Butterworth filter's pole locations presented in Figs. 4-3 through 4-9 were computed using the BASIC program given in Listing 4A-1. The first- and second-order stage coefficients listed in Tables 4-1 were computed using the BASIC program given in Listing 4A-2.

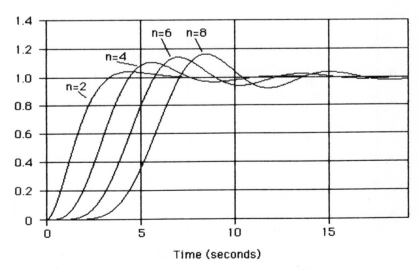

Fig. 4-19. Step response of even-order lowpass Butterworth filters.

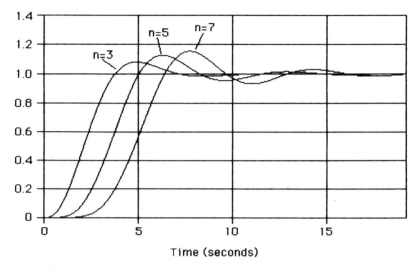

Fig. 4-20. Step response of odd-order lowpass Butterworth filters.

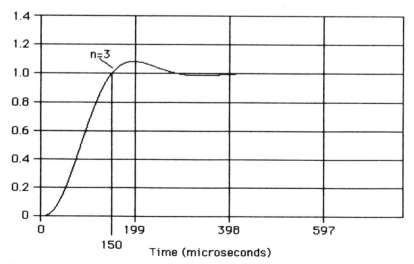

Fig. 4-21. Denormalized step response for Example 4-4.

COMPUTATION OF BUTTERWORTH
FREQUENCY RESPONSE DATA

The Butterworth magnitude responses shown in Figs. 4-10 and 4-11 were plotted using the BASIC program given in Listing 4B-1. The phase response shown in Fig. 4-12 were plotted using the BASIC program given in Listing 4B-2.

COMPUTATION OF TRANSIENT
RESPONSE DATA FOR BUTTERWORTH FILTERS

The transfer function for an n-th order lowpass Butterworth filter can be easily obtained by using Equations 4C-1 and 4C-2 as we saw in Section 4-1. To obtain the impulse response we need to take the inverse Laplace transform of the transfer function. The Heaviside expansion can be applied to the lowpass Butterworth case to produce Equations 4C-3 and 4C-4. The values of both K_r and s_r are in general complex, but for the lowpass Butterworth case all the complex pole values occur in complex conjugate pairs. When n is even this will allow Equation 4C-3 to be put in the form of Equation 4C-5. If n is odd there will be an additional pole at $s = -1$ which will cause the impulse response to be as given by Equation 4C-6 (Fig. 4-22). Although they look messy, Equations 4C-5 and 4C-6 are straightforward to implement in a BASIC program such as shown in Listing 4C-1. The step response is obtained via simple numerical integration of the impulse response.

$$H(s) = \frac{1}{(s-s_1)(s-s_2)\ldots(s-s_n)} \qquad \text{(Eq. 4C-1)}$$

$$s_i = e^{j\pi[(2i+n-1)/2n]} \qquad \text{(Eq. 4C-2)}$$

$$h(t) = \mathcal{L}^{-1}\left[h(t)\right] = \sum_{r=1}^{n} K_r \, e^{s_r t} \qquad \text{(Eq. 4C-3)}$$

$$K_r = \frac{(s-s_r)}{(s-s_1)(s-s_2)\ldots(s-s_n)}\bigg|_{s=s_r} \qquad \text{(Eq. 4C-4)}$$

$$h(t) = \sum_{r=1}^{n/2}\left[\, 2\,\text{Re}\{K_r\}\, e^{\sigma_r t} \cos(\omega_r t)\right.$$

$$\left. -\, 2\,\text{Im}\{K_r\}\, e^{\sigma_r t} \sin(\omega_r t)\right] \qquad \text{(Eq. 4C-5)}$$

$$h(t) = K\, e^{-t} + \sum_{r=1}^{n/2}\left[\, 2\,\text{Re}\{K_r\}\, e^{\sigma_r t} \cos(\omega_r t)\right.$$

$$\left. -\, 2\,\text{Im}\{K_r\}\, e^{\sigma_r t} \sin(\omega_r t)\right] \qquad \text{(Eq. 4C-6)}$$

Fig. 4-22. Formulas for computation of the Butterworth impulse response.

Listing 4A-1. BASIC program for computing Butterworth pole locations.

```
'*************************************************************
'
'   This program computes the pole locations for Butterworth filters
'   of orders 2 through 8. To use for other orders, change the range of k%
'   in the FOR NEXT loop. For hard copy output, change all PRINT commands
'   to LPRINT commands.
'
'*************************************************************
'
pi#= 3.1415926535898#
'
PRINT CHR$(12)
PRINT "Pole locations for Butterworth filters"
PRINT " "
PRINT " "
'
FOR n%=2 TO 8
PRINT "Order = ";n%
'
FOR k%=1 TO n%
r#=COS(pi#*(2*k%+n%-1)/(2*n%))
i#=SIN(pi#*(2*k%+n%-1)/(2*n%))
IF r#<= 0 AND i#>=0 THEN theta#=pi#-ATN(i#/ABS(r#))
IF r#<=0 AND i#<=0 THEN theta#=pi#+ATN(ABS(i#)/ABS(r#))
angle#=180!*theta#/pi#
PRINT USING "S[#] = ##.###### _+ j ##.######   ";k%;r#;i#;
PRINT USING"theta = ##.##### rad ( ###.# deg)";theta#;angle#
NEXT k%
'
PRINT " "
NEXT n%
STOP
```

Listing 4A-2. BASIC program for computing Butterworth filter coefficients.

```
'****************************************************************
'
'   This program computes coefficients for Butterworth filters
'   of orders 2 through 8.  To use for other orders, change the range of k%
'   in the FOR NEXT loop. For hard copy output, change all PRINT commands
'   to LPRINT commands.
'
'****************************************************************
'
pi#= 3.1415926535898#
'
PRINT CHR$(12)
PRINT "Coefficients for Butterworth filters realized by cascading"
PRINT "    first- and second-order stages"
PRINT " "
PRINT " "
'
FOR n%=2 TO 8
PRINT "Order = ";n%
'
FOR k%=1 TO n%\2
r#=COS(pi#*(2*k%+n%-1)/(2*n%))
i#=SIN(pi#*(2*k%+n%-1)/(2*n%))
b#=2!*ABS(r#)
'
PRINT USING "   b[#] = #.######   ";k%;b#
NEXT k%
'
PRINT " "
NEXT n%
STOP
```

Listing 4B-1. BASIC program for computing Butterworth magnitude response.

```
'********************************************************

OPTION BASE 0
DIM yval(384)
CLS
INPUT;"order of filter";order%
'
try.again:
INPUT " passband or stopband? (P/S)",band$
'
IF UCASE$(band$)<>"P" AND UCASE$(band$)<>"S" GOTO try.again
IF UCASE$(band$)="P" THEN CALL passband.setup
IF UCASE$(band$)="S" THEN CALL stopband.setup
'
CLS
CALL one.cyc.semi.log.box(ylab.val,ylab.incr)
'
e# = 2.7182818284592#
m# = 1
pi#=3.1415926535898#
'********************************************************
'  deltax% determines the horizontal spacing of computed plot points
'  set to 1 for smoothest plots, set to higher values for faster plots
deltax%=1
'
'********************************************************
'
CALL MOVETO(5,13)
PRINT "Butterworth ";band$;" response for order =";order%
IF band$="stopband" THEN GOTO stopband.proc
'
CALL butterworth.magnitude(yval(),deltax%,order%,1,1)
CALL magnitude.plot(yval(),deltax%,-3.5,.5,1)
CALL dynamic.halt
```

```
stopband.proc:
CALL butterworth.magnitude(yval(),deltax%,order%,10,1)
CALL magnitude.plot(yval(),deltax%,-80!,0!,1)
CALL dynamic.halt
'
'**********************************************
'   SUBROUTINES
'**********************************************
'
'   The following subprograms setup values needed for either passband
'   or stopband plotting.
'
SUB passband.setup STATIC
SHARED band$,ylab.val,ylab.incr
band$="passband"
ylab.val=.5
ylab.incr=.5
END SUB
'
SUB stopband.setup STATIC
SHARED band$,ylab.val,ylab.incr
band$="stopband"
ylab.val=0
ylab.incr=10
END SUB
'
'********************************************************
'   The following set of subprograms plots a semi-logarithmic
'   grid, one cycle by eight divisions.
'
SUB one.cyc.semi.log.box(ylab.val,ylab.incr) STATIC
CALL box.384.by.224
CALL vertical.grid.line(166,7)
CALL vertical.grid.line(233,7)
CALL vertical.grid.line(281,7)
CALL vertical.grid.line(318,7)
CALL vertical.grid.line(349,7)
CALL vertical.grid.line(375,7)
```

```
CALL vertical.grid.line(397,7)
CALL vertical.grid.line(416,7)
FOR iy%=48 TO 216 STEP 28
CALL horizontal.grid.line(iy%,7)
CALL MOVETO(440,iy%+5)
ylab.val=ylab.val-ylab.incr
PRINT USING "###.#";ylab.val
NEXT iy%
END SUB
'

SUB box.384.by.224 STATIC
LINE (50,20)-(50,244)
LINE (50,20)-(434,20)
LINE(50,244)-(434,244)
LINE(434,20)-(434,244)
END SUB
'

SUB vertical.grid.line(ix%,dot.interval%) STATIC
FOR iy%=20 TO 244 STEP dot.interval%
IF (iy%-20) MOD 28 <>0 THEN PSET(ix%,iy%)
NEXT iy%
END SUB
'

SUB horizontal.grid.line(iy%,dot.interval%) STATIC
FOR ix%=50 TO 434 STEP dot.interval%
PSET (ix%,iy%)
NEXT ix%
END SUB
'

'***********************************************************
'***********************************************************
'

'   The following subprogram plots the magnitude response contained in
'   the vector yval( ).
'

SUB magnitude.plot(yval(1),deltax%,ymin,ymax,tracetype%) STATIC
yrange=ymax-ymin
iyold%=244-224*((yval(0)-ymin)/yrange)
iyold2%=INT(244!-224!*((yval(0)-ymin)/yrange))
```

```
ixold%=0
FOR ix%=0 TO 383 STEP deltax%
iy%=244-224*((yval(ix%) - ymin)/yrange)
iy2%=INT(244!-224!*((yval(ix%)-ymin)/yrange))
IF iy%>244 THEN GOTO eof.pb.plot.loop
IF tracetype%=2 THEN PSET(ix%+50,iy%)
IF tracetype%=1 THEN LINE(ixold%+50,iyold%)-(ix%+50,iy%)
IF tracetype%=3 THEN LINE(ixold%+50,iyold2%)-(ix%+50,iy2%)
IF tracetype%=3 THEN LINE(ixold%+50,iyold2%+1)-(ix%+50,iy2%+1)
ixold%=ix%
iyold%=iy%
iyold2%=iy2%
eof.pb.plot.loop:
NEXT ix%
END SUB
'

'*********************************************************
'

'  The following subprogram computes the Butterworth magnitude response
'  for a given amount of passband ripple. The number of frequency
'  decades is given by freqcyc%. The order is given by n%.
'

SUB butterworth.magnitude(yval(1),deltax%,n%,maxfreq%,freqcyc%) STATIC
e# = 2.7182818284592#
m# = 1
pi#=3.1415926535898#
maxfreqexp%=LOG(maxfreq%)/2.302585093#
FOR ix%=0 TO 383 STEP deltax%
ss = (10^(maxfreqexp%+freqcyc%*(ix%-384)/384))
rp#=1
ip#=0
FOR k%=1 TO n%
x#=pi#*(n% + (2*k%)-1)/(2*n%)
r#=COS(x#)
i#=SIN(x#)
rpt#=ip#*(i#-ss)-rp#*r#
ipt#=rp#*(ss-i#)-r#*ip#
ip#=ipt#
rp#=rpt#
```

86

```
NEXT k%
p#=-ATN(ip#/rp#)
m#=1/(SQR(ip#*ip# + rp#*rp#))
yval(ix%)=20*LOG(m#)/2.302585093#
NEXT ix%
END SUB
'
'************************************************************
'************************************************************
'

SUB dynamic.halt STATIC
BEEP
done.loop:
GOTO done.loop
END SUB
```

Listing 4B-2. BASIC program for computing Butterworth phase response.

```
'************************************************************
'

OPTION BASE 0
DIM yval(384)
'
e# = 2.7182818284592#
m# = 1
pi#=3.1415926535898#
'
CLS
CALL two.cyc.semi.log.box
'************************************************************
' deltax% determines the horizontal spacing of computed plot points
'  set to 1 for smoothest plots, set to higher values for faster plots
deltax%=1
'

'************************************************************
'

INPUT; "Order of filter";order%
proc.time%=20+order%*25
PRINT "        wait";proc.time%;"seconds for results"
```

```
'
CALL butterworth.phase(yval(),deltax%,order%,10,2)
CALL butterworth.phase.smoother(yval(),deltax%,1)
'
CALL response.plot(yval(),deltax%,-720!,0!,1)
'
CALL dynamic.halt
'***********************************************
'  SUBROUTINES
'***********************************************
'  The following set of subprograms plots a semi-logarithmic grid,
'  two cycles by eight divisions
'
SUB two.cyc.semi.log.box STATIC
CALL box.384.by.224
CALL two.cycle.freq.grid(7)
FOR iy%=48 TO 216 STEP 28
CALL horizontal.grid.line(iy%,7)
NEXT iy%
END SUB
'
SUB box.384.by.224 STATIC
LINE (50,20)-(50,248)
LINE (46,20)-(438,20)
LINE(46,244)-(438,244)
LINE(434,20)-(434,248)
END SUB
'
SUB two.cycle.freq.grid(dot.interval%) STATIC
DIM two.cycle.freq.tics%(17)
two.cycle.freq.tics%(0)=108
two.cycle.freq.tics%(1)=141
two.cycle.freq.tics%(2)=166
two.cycle.freq.tics%(3)=184
two.cycle.freq.tics%(4)=200
two.cycle.freq.tics%(5)=213
two.cycle.freq.tics%(6)=224
two.cycle.freq.tics%(7)=233
two.cycle.freq.tics%(8)=242
two.cycle.freq.tics%(9)=300
```

```
two.cycle.freq.tics%(10)=333
two.cycle.freq.tics%(11)=356
two.cycle.freq.tics%(12)=376
two.cycle.freq.tics%(13)=392
two.cycle.freq.tics%(14)=405
two.cycle.freq.tics%(15)=416
two.cycle.freq.tics%(16)=425
FOR tic.index%=0 TO 16
ix%=two.cycle.freq.tics%(tic.index%)
FOR iy%=20 TO 244 STEP dot.interval%
IF (iy%-20) MOD 28 <>0 THEN PSET(ix%,iy%)
NEXT iy%
LINE(ix%,244)-(ix%,248)
NEXT tic.index%
LINE(242,20)-(242,244)
END SUB
'

SUB horizontal.grid.line(iy%,dot.interval%) STATIC
LINE(46,iy%)-(50,iy%)
FOR ix%=50 TO 434 STEP dot.interval%
PSET (ix%,iy%)
NEXT ix%
LINE(434,iy%)-(438,iy%)
END SUB
'

'**********************************************************
'

'   The following subprogram computes the Butterworth phase response.
'

SUB butterworth.phase(yval(1),deltax%,n%,maxfreq%,freqcyc%) STATIC
e# = 2.7182818284592#
m# = 1
pi#=3.1415926535898#
maxfreqexp%=LOG(maxfreq%)/2.302585093#
FOR ix%=0 TO 383 STEP deltax%
ss = (10^(maxfreqexp%+freqcyc%*(ix%-384)/384))
REM LPRINT ix%,ss
rp#=1
ip#=0
FOR k%=1 TO n%
```

```
x#=pi#*(n% + (2*k%)-1)/(2*n%)
r#=COS(x#)
i#=SIN(x#)
rpt#=ip#*(i#-ss)-rp#*r#
ipt#=rp#*(ss-i#)-r#*ip#
ip#=ipt#
rp#=rpt#
NEXT k%
p#=-ATN(ip#/rp#)
m#=1/(SQR(ip#*ip# + rp#*rp#))
yval(ix%)=p#*180#/pi#
NEXT ix%
END SUB
'
'*********************************************************
'
'   The following subprogram removes discontinuities in the phase
'   response arising from the use of only the principal value for
'   the arctan function.
'
SUB butterworth.phase.smoother(yval(1),deltax%,angletype%) STATIC
full.circle.offset=0
FOR ix%=deltax% TO 383 STEP deltax%
yval(ix%)=yval(ix%)+full.circle.offset
IF ABS(yval(ix%-deltax%)-yval(ix%))<90! THEN GOTO offset.ok
IF yval(ix%-deltax%)>yval(ix%) THEN GOTO increase.offset
yval(ix%)=yval(ix%)-180
full.circle.offset=full.circle.offset-180
GOTO offset.ok
increase.offset:
yval(ix%)=yval(ix%)+180
full.circle.offset=full.circle.offset+180
offset.ok:
NEXT ix%
END SUB
'
'*********************************************************
'
'   The following subprogram plots the phase response contained in
'   the vector yval().
```

```
SUB response.plot(yval(1),deltax%,ymin,ymax,tracetype%) STATIC
yrange=ymax-ymin
iyold%=244-224*((yval(0)-ymin)/yrange)
iyold2%=INT(244!-224!*((yval(0)-ymin)/yrange))
ixold%=0
FOR ix%=0 TO 383 STEP deltax%
iy%=244-224*((yval(ix%) - ymin)/yrange)
iy2%=INT(244!-224!*((yval(ix%)-ymin)/yrange))
IF iy%>244 THEN GOTO eof.pb.plot.loop
IF tracetype%=2 THEN PSET(ix%+50,iy%)
IF tracetype%=1 THEN LINE(ixold%+50,iyold%)-(ix%+50,iy%)
IF tracetype%=3 THEN LINE(ixold%+50,iyold2%)-(ix%+50,iy2%)
IF tracetype%=3 THEN LINE(ixold%+50,iyold2%+1)-(ix%+50,iy2%+1)
ixold%=ix%
iyold%=iy%
iyold2%=iy2%
eof.pb.plot.loop:
NEXT ix%
END SUB
'
'*****************************************************************
'
SUB dynamic.halt STATIC
BEEP
done.loop:
GOTO done.loop
END SUB
```

Listing 4C-1. BASIC program for computing Butterworth transient responses.

```
'*****************************************************************
'
'   This program computes the impulse and step responses for
'   Butterworth filters.
'
OPTION BASE 0
DIM yval(384)
'
```

```
deltax%=1
pi#=3.1415926535898#
'
INPUT;"order of Butterworth filter";n%
INPUT;"impulse or step response? (I/S) ";trans.type$
IF UCASE$(trans.type$)<>"I" AND UCASE$(trans.type$)<>"S" GOTO try.again
IF UCASE$(trans.type$)="S" THEN GOTO step.job
'
impulse.job:
CLS
PRINT USING"Lowpass Butterworth impulse response.  order=##";n%
CALL impulse.box
CALL butterworth.impulse(yval(),n%,deltax%,pi#)
CALL impulse.plot(yval(),0!,.7)
CALL dynamic.halt
'
step.job:
CLS
PRINT USING"Lowpass Butterworth step response.   order=##";n%
CALL impulse.box
CALL butterworth.impulse(yval(),n%,deltax%,pi#)
CALL integrate(yval(),yval.scaler)
CALL step.plot(yval(),0!,1.4,yval.scaler)
CALL dynamic.halt
'
'***********************************************************
'  SUBROUTINES
'***********************************************************
'
SUB dynamic.halt STATIC
BEEP
done.loop:
GOTO done.loop
END SUB
'
'***********************************************************
'
'  The following set of subprograms plots the grid for the impulse response.
'
SUB impulse.box STATIC
```

```
CALL box.384.by.224
CALL imp.grid
END SUB
'

SUB box.384.by.224 STATIC
LINE(50,20)-(50,244)
LINE(50,20)-(434,20)
LINE(50,244)-(434,244)
LINE(434,20)-(434,244)
END SUB
'

SUB imp.grid STATIC
LINE(50,212)-(434,212)
LINE(50,180)-(434,180)
LINE(50,148)-(434,148)
LINE(50,116)-(434,116)
LINE(50,84)-(434,84)
LINE(50,52)-(434,52)
LINE(150,20)-(150,244)
LINE(250,20)-(250,244)
LINE(350,20)-(350,244)
END SUB
'
'****************************************************************
'
'   The following subprogram plots the impulse response data
'   contained in the vector yval().
'
SUB impulse.plot(yval(1),ymin,yrange) STATIC
iyold%=244-224*((yval(0)-ymin)/yrange)
ixold%=0
'

FOR ix%=1 TO 383
iy%=244-224*((yval(ix%)+.2)/yrange)
IF iy%>244 THEN GOTO eof.plot.loop
LINE(ixold%+50,iyold%)-(ix%+50,iy%)
ixold%=ix%
iyold%=iy%
eof plot.loop:
NEXT ix%
'
```

```
END SUB
'
'****************************************************
'   The following subprogram integrates the impulse response to
'   obtain the step response.
'
SUB integrate(yval(1),yval.scaler) STATIC
'
FOR ix2%=1 TO 383
yval(ix2%)=yval(ix2%-1)+yval(ix2%)
NEXT ix2%
IF n%<4 THEN yval.scaler=yval(382)
END SUB
'
'****************************************************
'
'   The following subprogram plots the step response data contained
'   in the vector yval().
'
SUB step.plot(yval(1),ymin,yrange,yval.scaler) STATIC
'
iyold%=244-224*((yval(0)/yval.scaler-ymin)/yrange)
ixold%=0
'
FOR iix%=1 TO 383
iy%=244-224*((yval(iix%)/yval.scaler)/yrange)
IF iy%>244 THEN GOTO eof.plot.loop
REM PSET(iix%+50,iy%)
LINE(ixold%+50,iyold%)-(iix%+50,iy%)
ixold%=iix%
iyold%=iy%
eof.plot.loop:
NEXT iix%
END SUB
'
'*********************************************************
'
'   The following subprogram computes the impulse response of
'   a Butterworth filter of order n%.
'
```

94

```
SUB butterworth.impulse(yval(1),n%,deltax%,pi#) STATIC
FOR ix%=0 TO 383 STEP deltax%
CALL MOVETO(380,13)
PRINT 383-ix%
h.of.t#=0!
t=.05*ix%
FOR r%=1 TO n%\2
x#=pi# * (n% + (2*r%) -1) / (2 *n%)
sigma=COS(x#)
omega=SIN(x#)
'
' Compute Lr and Mr
'
L#=1
M#=0
'
FOR ii%=1 TO n%
IF ii%=r% GOTO end.I.loop
x#=pi# * (n% + (2* ii%) - 1) / (2 * n%)
R#= sigma - COS(x#)
I#= omega - SIN(x#)
'
LT#=L# * R# - M# * I#
MT#= L# * I# + R# * M#
L# = LT#
M# = MT#
end.I.loop:
NEXT ii%
'
L# = LT# / (LT# * LT# + MT# *MT#)
M# = -MT# / (LT# *LT# + MT# * MT#)
'
cos.part# = 2*L#*EXP(sigma*t)*COS(omega*t)
sin.part# = 2*M#*EXP(sigma*t)*SIN(omega*t)
'
h.of.t# = h.of.t# + cos.part# - sin.part#
NEXT r%
'
IF (n% MOD 2) = 0 THEN GOTO end.T.loop
'
```

```
'*********************************************************
' The following chunk of code computes the real exponential component
' present in odd-order responses.
'
'
K#=1
L#=1
M#=0
r%=(n%+1)/2
x#=pi# * (n% + (2*r%) -1) / (2 *n%)
sigma=COS(x#)
omega=SIN(x#)
FOR iii%=1 TO n%
IF iii%=r% GOTO end.III.loop
x#=pi# * (n% + (2* iii%) - 1) / (2 * n%)
R#= sigma - COS(x#)
I#= omega - SIN(x#)
'
LT#=L# * R# - M# * I#
MT#= L# * I# + R# * M#
L# = LT#
M# = MT#
'
end.III.loop:
NEXT iii%
K# = LT# / (LT# * LT# + MT# *MT#)
'
h.of.t# = h.of.t# + K#*EXP(-t)
'
'*********************************************************
end.T.loop:
'
yval(ix%)=h.of.t#
IF yval(ix%)>ymax THEN ymax=yval(ix%)
IF yval(ix%)<ymin THEN ymin=yval(ix%)
'
NEXT ix%
CALL MOVETO(380,13)
PRINT "    "
END SUB
```

Chapter 5

Chebyshev Filters

===

C HEBYSHEV LOWPASS FILTERS ARE DESIGNED TO HAVE AN amplitude response characteristic that has a relatively sharp transition from the passband to the stopband. This is accomplished at the expense of ripples that are introduced into the passband. As we will discover, the magnitude of these ripples can be controlled in the design process.

5.1 TRANSFER FUNCTION

The general shape of the Chebyshev magnitude response will be as shown in Fig. 5-1. This response can be normalized as in Fig. 5-2 so that the ripple bandwidth ω_r is equal to one. The response can also be normalized as in Fig. 5-3 so that the 3 dB frequency ω_0 is equal to one. Normalization based on the ripple bandwidth is simpler, but normalization based on the 3 dB point makes it easier to compare Chebyshev responses to those of other filter types.

The general expression for the transfer function of an n-th order Chebyshev lowpass filter is given by Equation 5.1-1 of Fig. 5-4. The poles are determined by Equations 5.1-3 through 5.1-7. Notice that this is somewhat more complicated than the pole formulas for Butterworth filters presented in Section 4.1, and several

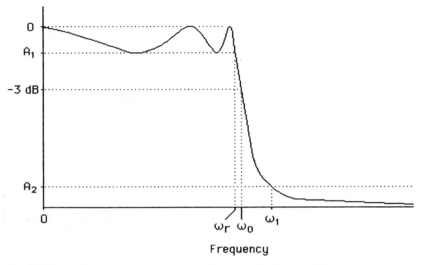

Fig. 5-1. Magnitude response of a typical lowpass Chebyshev filter.

parameters—ϵ, γ, and r—must be determined before the pole values can be calculated. Also the numerator must be calculated using all of the pole values.

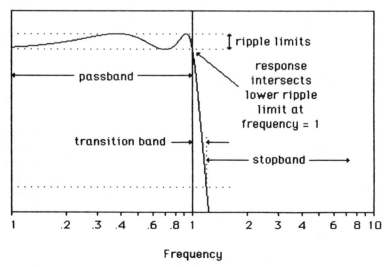

Fig. 5-2. Chebyshev response normalized to have passband end at $f = 1$.

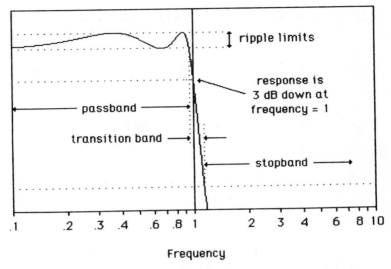

Fig. 5-3. Chebyshev response normalized to have 3 dB point at f = 1.

Design Procedure 5-1. To compute the poles of an n-th order Chebyshev lowpass filter normalized for a ripple bandwidth of one Hertz, perform the following steps:

1. Determine the maximum amount (in dB) of ripple which can be permitted in the passband magnitude response. Set r equal to or less than this value.
2. Use Equation 5.1-7 to compute ϵ.
3. Select an order n for the filter which will insure adequate performance.
4. Use Equation 5.1-6 to compute γ.
5. For i = 1, 2, 3 . . . n; use Equations 5.1-4 and 5.1-5 to compute the real part (σ_i) and imaginary part (ω_i) of each pole.
6. Use Equation 5.1-2 to compute H_o.
7. Substitute the values of H_o and s_1 through s_n into Equation 5.1-1.

Example 5-1. Determine the transfer function (normalized for ripple bandwidth equal to one) for a third-order Chebyshev filter with 0.5 dB passband ripple.

Solution. Figure 5-5 shows the steps involved in using Design Procedure 5-1 to determine the desired transfer function.

$$H(s) = \frac{H_0}{\prod\limits_{i=1}^{n} (s-s_i)} = \frac{H_0}{(s-s_1)(s-s_2)\ldots(s-s_n)}$$

$$(Eq.5.1-1)$$

$$H_0 = \begin{cases} \prod\limits_{i=1}^{n} (-s_i) & n \text{ odd} \\[4mm] 10^{r/20} \prod\limits_{i=1}^{n} (-s_i) & n \text{ even} \end{cases}$$

$$(Eq.5.1-2)$$

$$s_i = \sigma_i + j\omega_i \qquad (Eq.5.1-3)$$

$$\sigma_i = \left(\frac{\frac{1}{\gamma} - \gamma}{2} \right) \sin \frac{(2i-1)\pi}{2n} \qquad (Eq.5.1-4)$$

$$\omega_i = \left(\frac{\frac{1}{\gamma} + \gamma}{2} \right) \cos \frac{(2i-1)\pi}{2n} \qquad (Eq.5.1-5)$$

$$\gamma = \left(\frac{1 + \sqrt{1+\epsilon^2}}{\epsilon} \right)^{\frac{1}{n}} \qquad (Eq.5.1-6)$$

$$\epsilon = \sqrt{10^{r/10} - 1} \qquad (Eq.5.1-7)$$

Fig. 5-4. Transfer function for a lowpass Chebyshev filter.

$$\epsilon = \sqrt{10^{(0.5/10)} - 1} = 0.349311$$

$$Y = \left(\frac{1 + \sqrt{1 + (0.349311)^2}}{0.349311} \right)^{\frac{1}{3}} = 1.806477$$

$$\sigma_1 = \left(\frac{\frac{1}{1.806477} - 1.806477}{2} \right) \sin \frac{((2)(1) - 1)\pi}{(2)(3)}$$

$$= -0.313228$$

$$\omega_1 = \left(\frac{\frac{1}{1.806477} + 1.806477}{2} \right) \cos \frac{((2)(1) - 1)\pi}{(2)(3)}$$

$$= 1.021928$$

$$s_1 = -0.313228 + 1.021928\,j$$

$$s_2 = -0.626457$$

$$s_3 = -0.313228 - 1.021928\,j$$

$$H_0 = 0.715695$$

Fig. 5-5. Equations for Example 5-1.

From the form of (5.1-1) you can see that an n-th order Chebyshev filter will always have n poles and no finite zeros. These poles will lie on the left half of an ellipse in the s plane. The major axis of the ellipse lies on the $j\omega$ axis and the minor axis lies on the σ axis. The dimensions of the ellipse and the locations of the poles

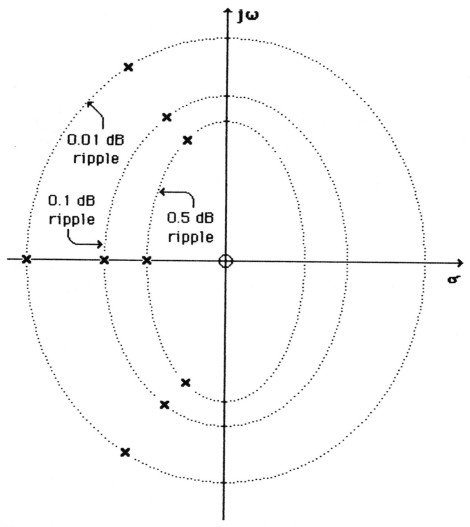

Fig. 5-6. Comparison of pole locations for third-order lowpass Chebyshev filters with different amounts of passband ripple.

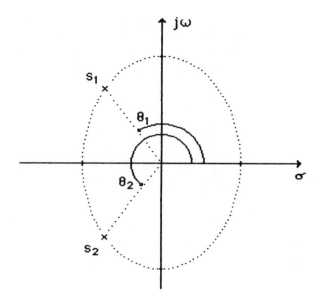

$$H(s) = \frac{H_0}{(s-s_1)(s-s_2)}$$

$s_1 = -0.712812 + 1.004042\,j$ $\theta_1 = 125.4° = 2.18816$ rad

$s_2 = -0.712812 - 1.004042\,j$ $\theta_2 = 234.6° = 4.09502$ rad

H = 1.516203

Fig. 5-7. Pole locations for a second-order lowpass Chebyshev filter with 0.5 dB passband ripple.

will depend upon the amount of ripple permitted in the passband. Values of passband ripple typically range from .01 dB to 1 dB. The smaller the passband ripple the wider the transition band will be. In fact for 0 dB ripple the Chebyshev filter and Butterworth filter have exactly the same transfer function and response characteristics. As the ripple limit is allowed to increase, the transition band will become narrower. For most moderately demanding applications, 0.5 dB of passband ripple represents a good compromise—the passband distortion due to the amplitude ripple will be tolerable, but the transition from passband to stopband will be significantly improved over that of a Butterworth filter of the same order. Pole locations for third-order Chebyshev filters having different ripple limits are compared in Fig. 5-6. Pole values, transfer functions,

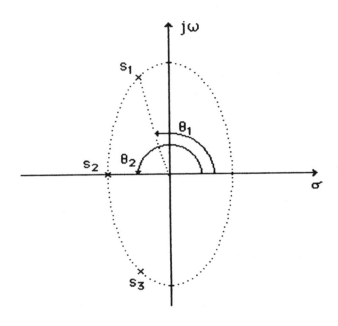

$$H(s) \quad = \quad \frac{H_0}{(s-s_1)(s-s_2)(s-s_3)}$$

$s_1 = -0.313228 + 1.021928 \, j$ $\qquad \theta_1 = 107.0° = 1.86821$ rad

$s_2 = -0.626457$ $\qquad \theta_2 = 180.0° = 3.14159$ rad

$s_3 = -0.313228 - 1.021928 \, j$ $\qquad \theta_3 = 253.0° = 4.41497$ rad

$H_0 = -0.715694$

Fig. 5-8. Pole locations for a third-order lowpass Chebyshev filter with 0.5 dB passband ripple.

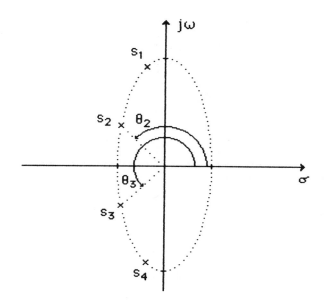

$$H(s) = \frac{H_0}{(s-s_1)(s-s_2)(s-s_3)(s-s_4)}$$

$s_1 = -0.175353 + 1.016253\ j$　　　　$\theta_1 = 99.8° = 1.74166$ rad

$s_2 = -0.423340 + 0.420946\ j$　　　　$\theta_2 = 135.2° = 2.35903$ rad

$s_3 = -0.423340 - 0.420946\ j$　　　　$\theta_3 = 224.8° = 3.92416$ rad

$s_4 = -0.175353 - 1.016253\ j$　　　　$\theta_4 = 260.2° = 4.54152$ rad

$H_0 = 0.379051$

Fig. 5-9. Pole locations for a fourth-order lowpass Chebyshev filter with 0.5 dB passband ripple.

and s-plane plots for 0.5 dB ripple Chebyshev filters of orders 2 through 8 are shown in Figs. 5-7 through 5-13. Pole values for various ripple limits are listed in Tables 5-1 through 5-3. (The BASIC programs used to generate these values are presented and explained at the end of the chapter.)

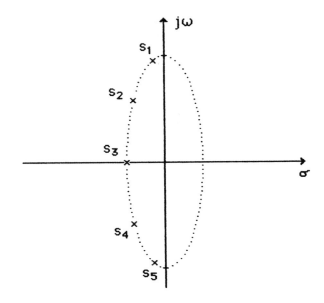

$$H(s) = \frac{H_0}{(s-s_1)(s-s_2)(s-s_3)(s-s_4)(s-s_5)}$$

$s_1 = -0.111963 + 1.011557 \, j$ $\theta_1 = 096.3° = 1.68103 \ \text{rad}$

$s_2 = -0.293123 + 0.625177 \, j$ $\theta_2 = 115.1° = 2.00923 \ \text{rad}$

$s_3 = -0.36232$ $\theta_3 = 180.0° = 3.14159 \ \text{rad}$

$s_4 = -0.293123 - 0.625177 \, j$ $\theta_4 = 244.9° = 4.27396 \ \text{rad}$

$s_5 = -0.111963 - 1.011557 \, j$ $\theta_5 = 263.7° = 4.60215 \ \text{rad}$

$H_0 = -0.178923$

Fig. 5-10. Pole locations for a fifth-order lowpass Chebyshev filter with 0.5 dB passband ripple.

The transfer function given by Equation 5.1-1 and the poles shown in the figures are normalized for a ripple bandwidth equal to one. To normalize for a 3 dB frequency of one, each pole s_i

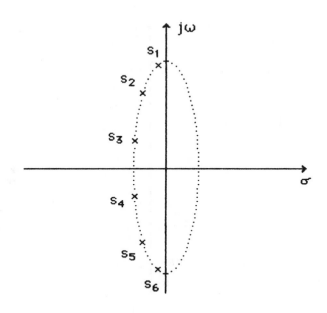

$$H(s) = \frac{H_0}{(s-s_1)(s-s_2)(s-s_3)(s-s_4)(s-s_5)(s-s_6)}$$

$s_1 = -0.077650 + 1.008461 \, j$ $\theta_1 = 094.4° = 1.64764$ rad

$s_2 = -0.212144 + 0.738245 \, j$ $\theta_2 = 106.0° = 1.85062$ rad

$s_3 = -0.289794 + 0.270216 \, j$ $\theta_3 = 137.0° = 2.39114$ rad

$s_4 = -0.289794 - 0.270216 \, j$ $\theta_4 = 223.0° = 3.89205$ rad

$s_5 = -0.212144 - 0.738245 \, j$ $\theta_5 = 254.0° = 4.43257$ rad

$s_6 = -0.077650 - 1.008461 \, j$ $\theta_6 = 265.6° = 4.63554$ rad

$H_0 = 0.094763$

Fig. 5-11. Pole locations for a sixth-order lowpass Chebyshev filter with 0.5 dB passband ripple.

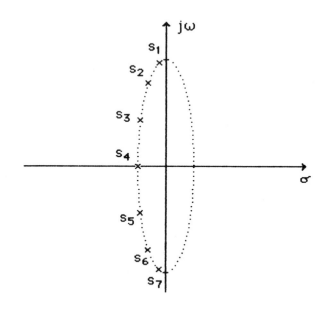

$$H(s) = \frac{H_0}{(s-s_1)(s-s_2)(s-s_3)(s-s_4)(s-s_5)(s-s_6)(s-s_7)}$$

$s_1 = -0.057003 + 1.006409\,j$ $\theta_1 = 093.2° = 1.62738$ rad

$s_2 = -0.159719 + 0.807077\,j$ $\theta_2 = 101.2° = 1.76617$ rad

$s_3 = -0.230801 + 0.447894\,j$ $\theta_3 = 117.3° = 2.04661$ rad

$s_4 = -0.25617$ $\theta_4 = 180.0° = 3.14159$ rad

$s_5 = -0.230801 - 0.447894\,j$ $\theta_5 = 242.7° = 4.23657$ rad

$s_6 = -0.159719 - 0.807077\,j$ $\theta_6 = 258.8° = 4.51701$ rad

$s_7 = -0.057003 - 1.006409\,j$ $\theta_7 = 266.8° = 4.65581$ rad

$H_0 = -0.044731$

Fig. 5-12. Pole locations for a seventh-order lowpass Chebyshev filter with 0.5 dB passband ripple.

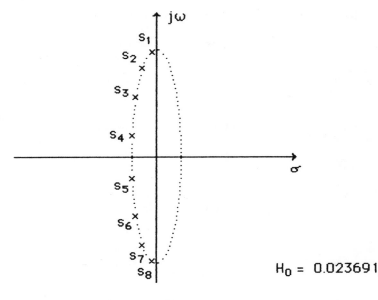

$$H(s) = \frac{H_0}{(s-s_1)(s-s_2)(s-s_3)(s-s_4)(s-s_5)(s-s_6)(s-s_7)(s-s_8)}$$

$s_1 = -0.043620 + 1.005002\,j$ $\theta_1 = 92.5° = 1.61417$ rad

$s_2 = -0.124219 + 0.852\,j$ $\theta_2 = 98.3° = 1.71557$ rad

$s_3 = -0.185908 + 0.569288\,j$ $\theta_3 = 108.1° = 1.88644$ rad

$s_4 = -0.219293 + 0.199907\,j$ $\theta_4 = 137.6° = 2.40241$ rad

$s_5 = -0.219293 - 0.199907\,j$ $\theta_5 = 222.4° = 3.88078$ rad

$s_6 = -0.185908 - 0.569288\,j$ $\theta_6 = 251.9° = 4.39675$ rad

$s_7 = -0.124219 - 0.852\,j$ $\theta_7 = 261.7° = 4.56761$ rad

$s_8 = -0.043620 - 1.005002\,j$ $\theta_8 = 267.5° = 4.66901$ rad

Fig. 5-13. Pole locations for an eighth-order lowpass Chebyshev filter with 0.5 dB passband ripple.

Table 5-1. Pole Values for Lowpass Chebyshev Filters with 0.1 dB Passband Ripple.

	Real part	Imaginary Part	Phase Angles Degrees	Radians
Order = 2	-1.186178	+1.380948	130.7	2.28047
H_0 = 3.314037	-1.186178	-1.380948	229.3	4.00272
Order = 3	-0.484703	+1.206155	111.9	1.95290
H_0 = -1.638051	-0.969406		180.0	3.14159
	-0.484703	-1.206155	248.1	4.33028
Order = 4	-0.264156	+1.122610	103.2	1.80190
H_0 = 0.828509	-0.637730	+0.465000	143.9	2.51157
	-0.637730	-0.465000	216.1	3.77162
	-0.264156	-1.122610	256.8	4.48129
Order = 5	-0.166534	+1.080372	98.8	1.72374
H_0 =-0.409513	-0.435991	+0.667707	123.1	2.14926
	-0.538914		180.0	3.14159
	-0.435991	-0.667707	236.9	4.13393
	-0.166534	-1.080372	261.2	4.55945
Order = 6	-0.114693	+1.056519	96.2	1.67893
H_0 = 0.207127	-0.313348	+0.773426	112.1	1.95573
	-0.428041	+0.283093	146.5	2.55727
	-0.428041	-0.283093	213.5	3.72592
	-0.313348	-0.773426	247.9	4.32746
	-0.114693	-1.056519	263.8	4.60425
Order = 7	-0.083841	+1.041833	94.6	1.65110
H_0 =-0.102378	-0.234917	+0.835485	105.7	1.84489
	-0.339465	+0.463659	126.2	2.20277
	-0.376778		180.0	3.14159
	-0.339465	-0.463659	233.8	4.08041
	-0.234917	-0.835485	254.3	4.43829
	-0.083841	-1.041833	265.4	4.63209
Order = 8	-0.063980	+1.032181	93.5	1.63270
H_0 = 0.051782	-0.182200	+0.875041	101.8	1.77608
	-0.272682	+0.584684	115.0	2.00718
	-0.321650	+0.205314	147.4	2.57348
	-0.321650	-0.205314	212.6	3.70971
	-0.272682	-0.584684	245.0	4.27600
	-0.182200	-0.875041	258.2	4.50710
	-0.063980	-1.032181	266.5	4.65048

Table 5-2. Pole Values for Lowpass Chebyshev Filters with 0.5 dB Passband Ripple.

	Real Part	Imaginary Part	Phase Angle Degrees	Phase Angle Radians
Order = 2	-0.712812	+1.004042	125.4	2.18816
$H_0 = 1.516203$	-0.712812	-1.004042	234.6	4.09502
Order = 3	-0.313228	+1.021928	107.0	1.86821
$H_0 = -0.715694$	-0.626457		180.0	3.14159
	-0.313228	-1.021928	253.0	4.41497
Order = 4	-0.175353	+1.016253	99.8	1.74166
$H_0 = 0.379051$	-0.423340	+0.420946	135.2	2.35903
	-0.423340	-0.420946	224.8	3.92416
	-0.175353	-1.016253	260.2	4.54152
Order = 5	-0.111963	+1.011557	96.3	1.68103
$H_0 = -0.178923$	-0.293123	+0.625177	115.1	2.00923
	-0.362320		180.0	3.14159
	-0.293123	-0.625177	244.9	4.27396
	-0.111963	-1.011557	263.7	4.60215
Order = 6	-0.077650	+1.008461	94.4	1.64764
$H_0 = 0.094763$	-0.212144	+0.738245	106.0	1.85062
	-0.289794	+0.270216	137.0	2.39114
	-0.289794	-0.270216	223.0	3.89205
	-0.212144	-0.738245	254.0	4.43257
	-0.077650	-1.008461	265.6	4.63554
Order = 7	-0.057003	+1.006409	93.2	1.62738
$H_0 = -0.044731$	-0.159719	+0.807077	101.2	1.76617
	-0.230801	+0.447894	117.3	2.04661
	-0.256170		180.0	3.14159
	-0.230801	-0.447894	242.7	4.23657
	-0.159719	-0.807077	258.8	4.51701
	-0.057003	-1.006409	266.8	4.65581
Order = 8	-0.043620	+1.005002	92.5	1.61417
$H_0 = 0.023691$	-0.124219	+0.852000	98.3	1.71557
	-0.185908	+0.569288	108.1	1.88644
	-0.219293	+0.199907	137.6	2.40241
	-0.219293	-0.199907	222.4	3.88078
	-0.185908	-0.569288	251.9	4.39675
	-0.124219	-0.852000	261.7	4.56761
	-0.043620	-1.005002	267.5	4.66901

Table 5-3. Pole Values for Lowpass Chebyshev Filters with 1.0 dB Passband Ripple.

	Real Part	Imaginary Part	Phase Angle Degrees	Phase Angle Radians
Order = 2	-0.548867	+0.895129	121.5	2.12084
H_0 = 1.102510	-0.548867	-0.895129	238.5	4.16234
Order = 3	-0.247085	+0.965999	104.3	1.82121
H_0 = -0.491307	-0.494171	-0.000000	180.0	3.14159
	-0.247085	-0.965999	255.7	4.46198
Order = 4	-0.139536	+0.983379	98.1	1.71175
H_0 = 0.275628	-0.336870	+0.407329	129.6	2.26180
	-0.336870	-0.407329	230.4	4.02139
	-0.139536	-0.983379	261.9	4.57144
Order = 5	-0.089458	+0.990107	95.2	1.66090
H_0 = -0.122827	-0.234205	+0.611920	110.9	1.93633
	-0.289493	-0.000000	180.0	3.14159
	-0.234205	-0.611920	249.1	4.34685
	-0.089458	-0.990107	264.8	4.62228
Order = 6	-0.062181	+0.993411	93.6	1.63331
H_0 = 0.068907	-0.169882	+0.727227	103.1	1.80028
	-0.232063	+0.266184	131.1	2.28782
	-0.232063	-0.266184	228.9	3.99537
	-0.169882	-0.727227	256.9	4.48290
	-0.062181	-0.993411	266.4	4.64988
Order = 7	-0.045709	+0.995284	92.6	1.61669
H_0 = -0.030707	-0.128074	+0.798156	99.1	1.72990
	-0.185072	+0.442943	112.7	1.96657
	-0.205414	-0.000000	180.0	3.14159
	-0.185072	-0.442943	247.3	4.31661
	-0.128074	-0.798156	260.9	4.55328
	-0.045709	-0.995284	267.4	4.66650
Order = 8	-0.035008	+0.996451	92.0	1.60591
H_0 = 0.017227	-0.099695	+0.844751	96.7	1.68827
	-0.149204	+0.564444	104.8	1.82922
	-0.175998	+0.198206	131.6	2.29692
	-0.175998	-0.198206	228.4	3.98627
	-0.149204	-0.564444	255.2	4.45396
	-0.099695	-0.844751	263.3	4.59492
	-0.035008	-0.996451	268.0	4.67727

must be divided a factor R and the numerator must be divided by R^n as shown in Equation 5.1-8 of Fig. 5-14.

Design Procedure 5-2. A Chebyshev lowpass filter normalized for a ripple bandwidth of one can be renormalized for a 3 dB frequency of one by performing the following steps.

$$H_{3dB}(s) = \frac{H_0/R^n}{\prod_{i=1}^{n}\left(s - \frac{s_i}{R}\right)} \qquad (Eq. 5.1-8)$$

$$R = \cosh A$$

$$= \frac{e^A + e^{-A}}{2} \qquad (Eq. 5.1-9)$$

$$A = \frac{\cosh^{-1}\left(\frac{1}{\epsilon}\right)}{n} \qquad (Eq. 5.1-10)$$

$$= \frac{1}{n} \log\left[\frac{1 + \sqrt{1 - \epsilon^2}}{\epsilon}\right]$$

Fig. 5-14. Formulas for Design Procedure 5-2.

1. Compute the factor A using Equation 5.1-10.
2. Using this value of A in Equation 5.1-9, compute the factor R. (Table 5-4 lists R factors for various orders and ripple limits. If the required combination can be found in this table, Steps 1 and 2 can be skipped.)
3. Divide each pole value s_i by R.
4. Divide the numerator H_o by R^n.

Higher-order filters are often implemented as cascaded first- and second-order sections. When designing these sections, it is very convenient to have the transfer function in a form in which each pair of complex conjugate poles is combined into a single second-order filter section. If the overall filter order is odd, the one real pole will stand alone as a first-order section. Tables 5-5 through 5-7 list the appropriate section coefficients for Chebyshev filters of orders 2 through 8 for passband ripple of 0.1 dB, 0.5 dB, and 1.0 dB. The filters formed from these sections are normalized for a 3 dB frequency of one.

5.2 FREQUENCY RESPONSE

Figures 5-15 through 5-26 show the passband amplitude response, stopband amplitude response, and phase response for

Table 5-4. Factors for Renormalizing Chebyshev Transfer Functions.

order →	2	3	4	5	6	7	8
ripple							
0.1	1.94322	1.38899	1.21310	1.13472	1.09293	1.06800	1.05193
0.2	1.67427	1.28346	1.15635	1.09915	1.06852	1.05019	1.03835
0.3	1.53936	1.22906	1.12680	1.08055	1.05571	1.04083	1.03121
0.4	1.45249	1.19348	1.10736	1.06828	1.04725	1.03464	1.02649
0.5	1.38974	1.16749	1.09310	1.05926	1.04103	1.03009	1.02301
0.6	1.34127	1.14724	1.08196	1.05220	1.03616	1.02652	1.02028
0.7	1.30214	1.13078	1.07288	1.04644	1.03218	1.02361	1.01806
0.8	1.26955	1.11699	1.06526	1.04160	1.02883	1.02116	1.01618
0.9	1.24176	1.10517	1.05872	1.03745	1.02596	1.01905	1.01457
1.0	1.21763	1.09487	1.05300	1.03381	1.02344	1.01721	1.01316
1.1	1.19637	1.08576	1.04794	1.03060	1.02121	1.01557	1.01191
1.2	1.17741	1.07761	1.04341	1.02771	1.01922	1.01411	1.01079
1.3	1.16035	1.07025	1.03931	1.02510	1.01741	1.01278	1.00978
1.4	1.14486	1.06355	1.03558	1.02272	1.01576	1.01157	1.00886
1.5	1.13069	1.05740	1.03216	1.02054	1.01425	1.01046	1.00801

**Table 5-5. Coefficients for 0.1 dB Chebyshev
Filters Realized as Cascaded First- and Second-Order Sections.**

	B_i	C_i
Order = 2		
	2.372356	3.314037
Order = 3		
	0.969406	1.689747
		0.969406
Order = 4		
	0.528313	1.330031
	1.275460	0.622925
Order = 5		
	0.333067	1.194937
	0.871982	0.635920
		0.538914
Order = 6		
	0.229387	1.129387
	0.626696	0.696374
	0.856083	0.263361
Order = 7		
	0.167682	1.092446
	0.469834	0.753222
	0.678930	0.330217
		0.376778
Order = 8		
	0.127960	1.069492
	0.364400	0.798894
	0.545363	0.416210
	0.643300	0.145612

Chebyshev filters of orders two through eight for various values
of passband ripple. (The BASIC program used to generate these
various response plots are presented and explained at the end of
the chapter.) These frequency response plots are normalized for
a cutoff frequency of 1 hertz. To denormalize them, simply multiply the frequency axis by the desired cutoff frequency f_c.

Table 5-6. Coefficients for 0.5 dB Chebyshev
Filters Realized as Cascaded First- and Second-Order Sections.

	B_i	C_i
Order = 2		
	1.425625	1.516203
Order = 3		
	0.626457	1.142448
		0.626457
Order = 4		
	0.350706	1.063519
	0.846680	0.356412
Order = 5		
	0.223926	1.035784
	0.586245	0.476767
		0.362320
Order = 6		
	0.155300	1.023023
	0.424288	0.590010
	0.579588	0.156997
Order = 7		
	0.114006	1.016108
	0.319439	0.676884
	0.461602	0.253878
		0.256170
Order = 8		
	0.087240	1.011932
	0.248439	0.741334
	0.371815	0.358650
	0.438586	0.088052

Example 5-2. Use Figs. 5-18 and 5-25 to determine the magnitude and phase response at 120 Hz of a sixth-order 0.5 dB Chebyshev lowpass filter having a cutoff frequency of 400 Hz.

Solution. If we set $f_c = 400$ you can denormalize the $n = 6$ response of Fig. 5-18 to obtain the response shown in Fig. 5-27. Then the magnitude response at 120 Hz is approximately -0.05 dB. Likewise, you can denormalize the $n = 6$ response of Fig. 5-25 to obtain the response shown in Fig. 5-28. From this the phase response at 120 Hz is approximately $-90°$.

5.3 IMPULSE RESPONSE

Impulse responses for lowpass Chebyshev filters are shown in Figs. 5-29 through 5-34. (The BASIC program used to generate

Table 5-7. Coefficients for 1.0 dB Chebyshev Filters Realized as Cascaded First- and Second-Order Sections.

	B_i	C_i
Order = 2		
	1.097734	1.102510
Order = 3		
	0.494171	0.994205
		0.494171
Order = 4		
	0.279072	0.986505
	0.673739	0.279398
Order = 5		
	0.178917	0.988315
	0.468410	0.429298
		0.289493
Order = 6		
	0.124362	0.990732
	0.339763	0.557720
	0.464125	0.124707
Order = 7		
	0.091418	0.992679
	0.256147	0.653456
	0.370144	0.230450
		0.205414
Order = 8		
	0.070016	0.994141
	0.199390	0.723543
	0.298408	0.340859
	0.351997	0.070261

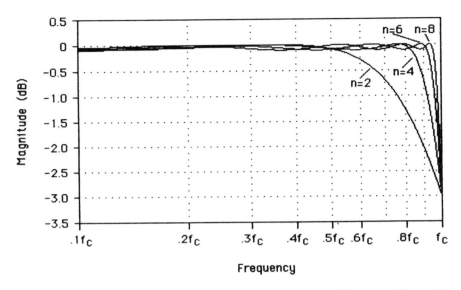

Fig. 5-15. Passband magnitude response of even-order lowpass Chebyshev filters with 0.1 dB passband ripple.

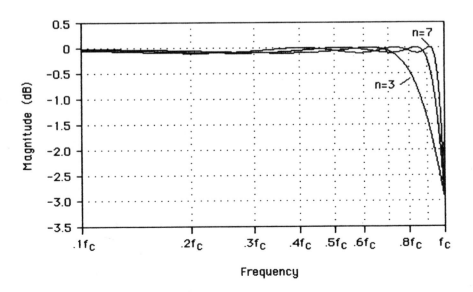

Fig. 5-16. Passband magnitude response of odd-order lowpass Chebyshev filters with 0.1 dB passband ripple.

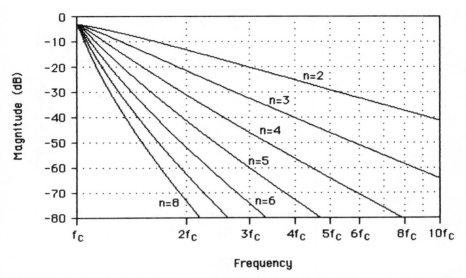

Fig. 5-17. Stopband magnitude response of lowpass Chebyshev filters with 0.1 dB passband ripple.

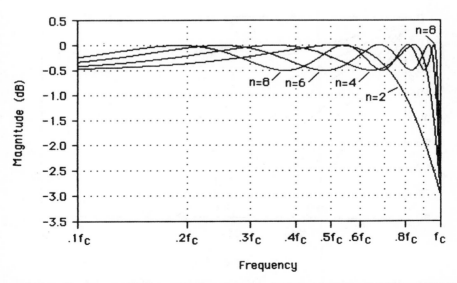

Fig. 5-18. Passband magnitude response of even-order lowpass Chebyshev filters with 0.5 dB passband ripple.

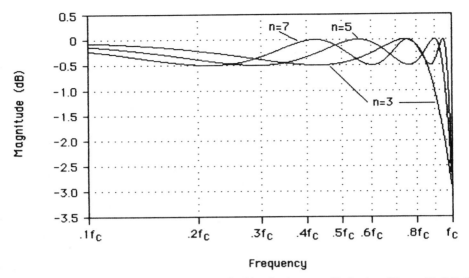

Fig. 5-19. Passband magnitude response of odd-order lowpass Chebyshev filters with 0.5 dB passband ripple.

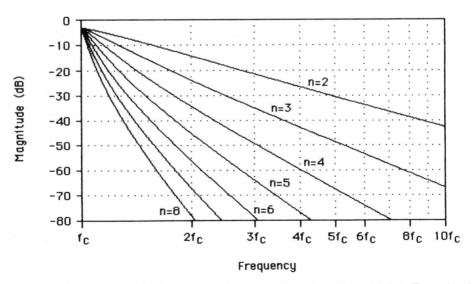

Fig. 5-20. Stopband magnitude response of lowpass Chebyshev filters with 0.5 dB passband ripple.

120

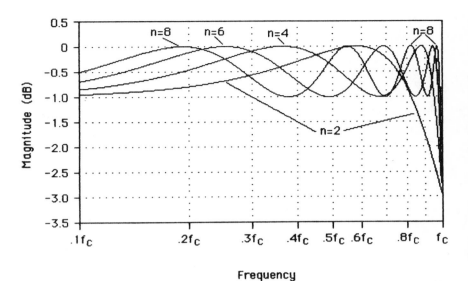

Fig. 5-21. Passband magnitude response of even-order lowpass Chebyshev filters with 1 dB passband ripple.

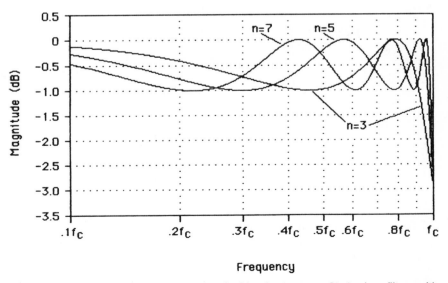

Fig. 5-22. Passband magnitude response of odd-order lowpass Chebyshev filters with 1 dB passband ripple.

121

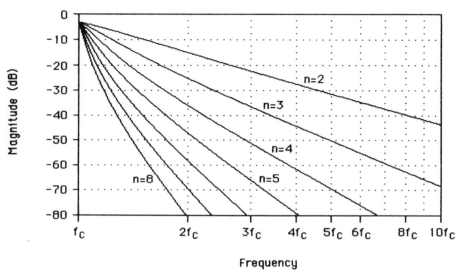

Fig. 5-23. Stopband magnitude response of lowpass Chebyshev filters with 1 dB passband ripple.

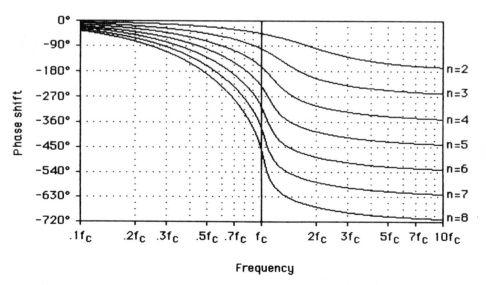

Fig. 5-24. Phase response of lowpass Chebyshev filters with 0.1 dB passband ripple.

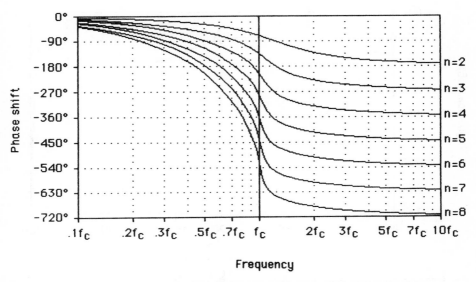

Fig. 5-25. Phase response of lowpass Chebyshev filters with 0.5 dB passband ripple.

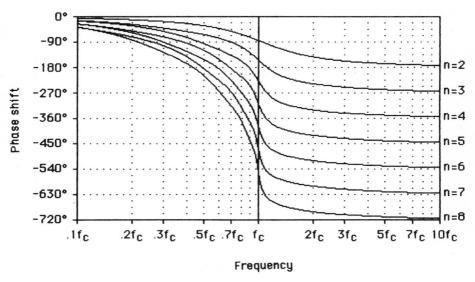

Fig. 5-26. Phase response of lowpass Chebyshev filters with 1 dB passband ripple.

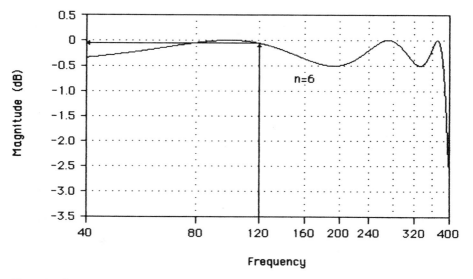

Fig. 5-27. Denormalized magnitude response for Example 5-2.

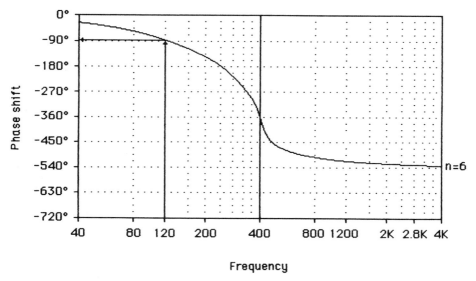

Fig. 5-28. Denormalized phase response for Example 5-2.

124

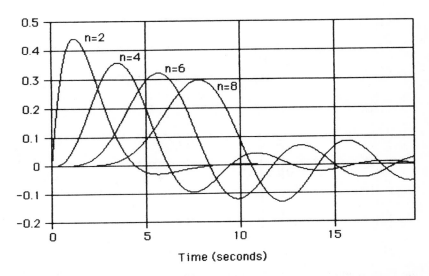

Fig. 5-29. Impulse response of even-order lowpass Chebyshev filters with 0.1 dB. passband ripple.

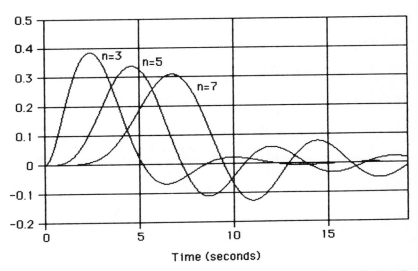

Fig. 5-30. Impulse response of odd-order lowpass Chebyshev filters with 0.1 dB passband ripple.

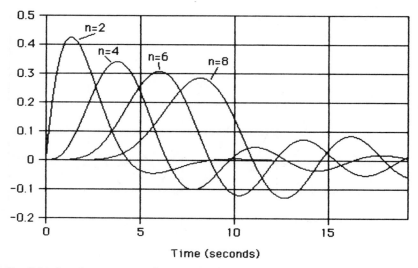

Fig. 5-31. Impulse response of even-order lowpass Chebyshev filters with 0.5 dB passband ripple.

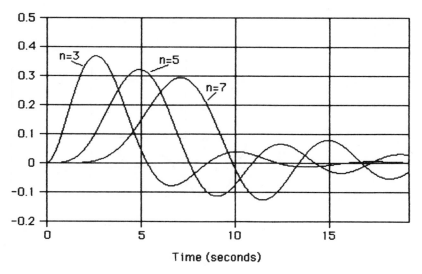

Fig. 5-32. Impulse response of odd-order lowpass Chebyshev filters with 0.5 dB passband ripple.

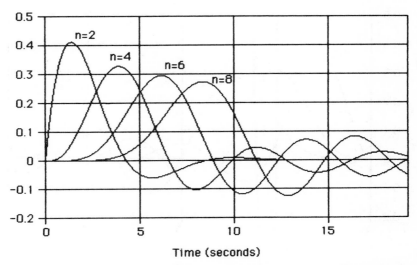

Fig. 5-33. Impulse response of even-order lowpass Chebyshev filters with 1.0 dB passband ripple.

these responses is presented and discussed at the end of the chapter.) These responses are normalized for lowpass filters having a cutoff frequency f_c equal to 1 Hz. To denormalize the response, divide the time axis by the desired cutoff frequency f_c and multiply the amplitude axis by the same factor.

Example 5-3. Use Fig. 5-34 to determine the instantaneous amplitude of the output 15 msec after a unit impulse is applied to the input of a fifth-order Chebyshev LPF having $f_c = 250$ Hz and passband ripple of 1 dB.

Solution. If we set $f_c = 250$ you can denormalize the $n = 5$ response of Fig. 5-34 to obtain the response shown in Fig. 5-35. From this you can observe that the response amplitude at $t = 15$ msec is approximately 56.

5.4 STEP RESPONSE

Step responses for lowpass Chebyshev filters are shown in Figs. 5-36 through 5-41. (The BASIC program used to generate these responses is presented and discussed at the end of the chapter.) These responses are normalized for lowpass filters having a cutoff frequency equal to 1 Hz. To denormalize the response, divide the time axis by the desired cutoff frequency f_c.

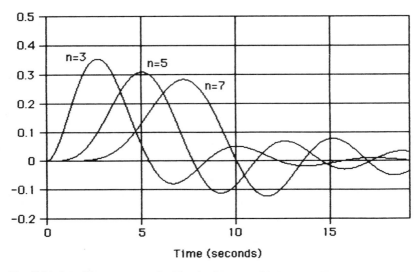

Fig. 5-34. Impulse response of odd-order lowpass Chebyshev filters with 1.0 dB passband ripple.

Example 5-4. Use Fig. 5-39 to determine how long it will take for the step response of a seventh-order Chebyshev LPF ($f_c = 4$ kHz, ripple = 0.5 dB) to first reach 100 percent of its final value.

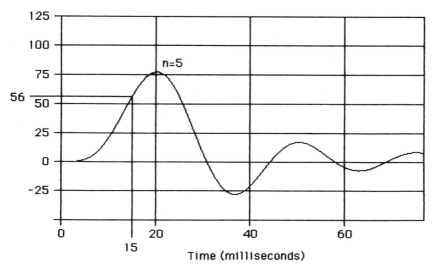

Fig. 5-35. Denormalized impulse response for Example 5-3.

128

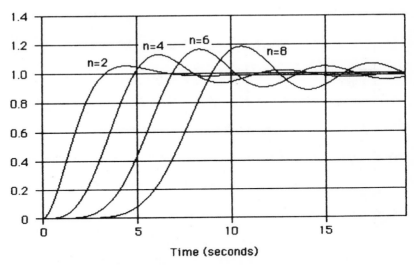

Fig. 5-36. Step response of even-order lowpass Chebyshev filters with 0.1 dB passband ripple.

Solution. If we set $f_c = 4000$, we can denormalize the $n = 7$ response of Fig. 5-39 to obtain the response shown in Fig. 5-42. From this you can observe that the time required for the step response to reach a value of one is approximately 2.1 milliseconds.

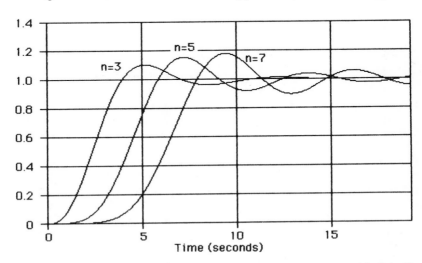

Fig. 5-37. Step response of odd-order lowpass Chebyshev filters with 0.1 dB passband ripple.

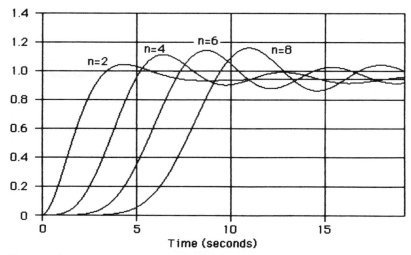

Fig. 5-38. Step response of even-order lowpass Chebyshev filters with 0.5 dB passband ripple.

COMPUTATION OF
CHEBYSHEV TRANSFER FUNCTION DATA

The Chebyshev filter's pole locations presented in Figs. 5-6 through 5-13 and Tables 5-1 through 5-3 were computed using the

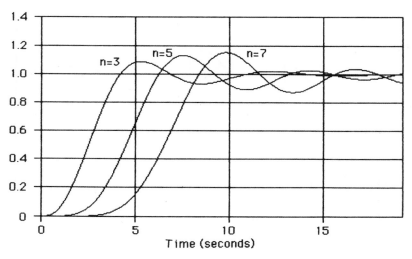

Fig. 5-39. Step response of odd-order lowpass Chebyshev filters with 0.5 dB passband ripple.

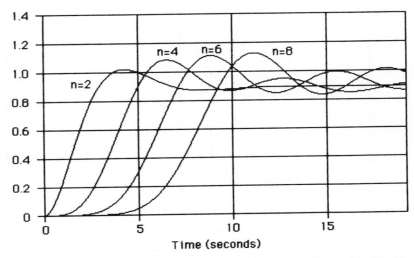

Fig. 5-40. Step response of even-order lowpass Chebyshev filters with 1.0 dB passband ripple.

BASIC program given in Listing 5A-1. The renormalization factors listed in Table 5-4 were computed using the BASIC program given in Listing 5A-2. The first- and second-order stage coefficients listed in Tables 5-5 through 5-7 were computed using the BASIC program given in Listing 5A-3.

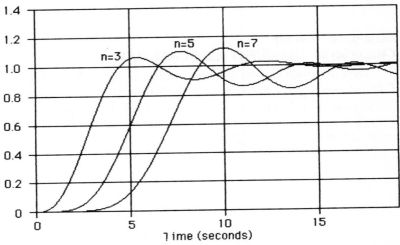

Fig. 5-41. Step response of odd-order lowpass Chebyshev filters with 1.0 dB passband ripple.

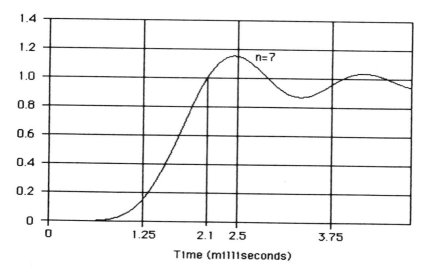

Fig. 5-42. Denormalized step response for Example 5-4.

COMPUTATION OF
CHEBYSHEV FREQUENCY RESPONSE DATA

The Chebyshev magnitude responses shown in Figs. 5-15 through 5-23 were plotted using the BASIC program given in Listing 5B-1. The phase responses shown in Figs. 5-24 through 5-26 were plotted using the BASIC program given in Listing 5B-2.

COMPUTATION OF
CHEBYSHEV TRANSIENT RESPONSE DATA

The Chebyshev impulse responses shown in Figs. 5-29 through 5-34 and step responses shown in Figs. 5-36 through 5-41 were plotted using the BASIC program given in Listing 5C-1. This program uses the Heaviside expansion as discussed for the Butterworth transient responses in Listing 4C.

Listing 5A-1. BASIC program for computing Chebyshev pole locations.

```
pi#=3.1415926535898#
CLS
LPRINT CHR$(12)
INPUT "Passband ripple in dB";ripple#
epsilon#=SQR(10^(ripple#/10!)-1)

FOR order%=2 TO 8
LPRINT USING"type = Cheby    ripple = ##.##   order = #";ripple#;order%
gamma# = ((1+SQR(1+epsilon#*epsilon#))/epsilon#)^(1!/order%)
neg.sinh.gamma#=(1!/gamma#-gamma#)/2!
cosh.gamma#=(1!/gamma#+gamma#)/2!
real.H.zero#=1!
imag.H.zero#=0!

FOR i%=1 TO order%
sigma#=neg.sinh.gamma#*SIN((2*i%-1)*pi#/(2*order%))
omega#=cosh.gamma#*COS((2*i%-1)*pi#/(2*order%))
theta.rad=pi#+ ATN(omega#/sigma#)
theta.deg=theta.rad*180!/pi#
LPRINT USING"s(#) = ##.###### + j  ##.######";i%;sigma#;omega#;
LPRINT USING"   theta = ###.# deg = #.##### rad";theta.deg;theta.rad
rpt#= real.H.zero#*sigma#-imag.H.zero#*omega#
ipt#= real.H.zero#*omega#+imag.H.zero#*sigma#
real.H.zero#=rpt#
imag.H.zero#=ipt#
NEXT i%

LPRINT USING"Ho = ##.###### + j  ##.######";real.H.zero#;imag.H.zero#
LPRINT " "
NEXT order%
STOP
```

Listing 5A-2. BASIC program for computing renormalization factors for Chebyshev filters.

```
'   This program computes factors for renormalizing Chebysev filters
'   from a ripple bandwidth of one to a 3 dB frequency of one
'
'**************************************************************
'
OPEN "O",#1,"SCRN:"
CLS
PRINT #1, CHR$(12)
'
FOR i%=1 TO 15
'
ripple#=i%*.1
'
epsilon#=SQR(10^(ripple#/10!)-1)
work#=1!/epsilon#
'
FOR order%=2 TO 8
'
A=(LOG(work#+SQR(work#*work#-1)))/order%
R(order%)=(EXP(A)+EXP(-A))/2
'
NEXT order%
PRINT #1, USING"#.#     #.#####  #.##### ";ripple#;R(2);R(3);
PRINT #1, USING"#.#####  #.#####  #.##### ";R(4);R(5);R(6);
PRINT #1, USING"#.#####  #.#####";R(7);R(8)
NEXT i%
'
STOP
```

Listing 5A-3. BASIC program for computing Chebyshev filter coefficients.

```
'
'   This program computes coefficients for Chebyshev filters
'   of orders 2 through 8.  To use for other orders, change the
'   range of k% in the FOR NEXT loop.
'
'**************************************************************
'
```

```
pi#= 3.1415926535898#
'
CLS
INPUT "Passband ripple in dB";ripple#
epsilon#=SQR(10^(ripple#/10!)-1)
PRINT CHR$(12)
PRINT "Coefficients for Chebyshev filters realized by cascading"
PRINT "    first- and second-order stages"
PRINT " "
PRINT USING "Passband ripple = #.# dB";ripple#
PRINT " "
'
FOR order%=2 TO 8
PRINT USING"order = #";order%
gamma# = ((1+SQR(1+epsilon#*epsilon#))/epsilon#)^(1!/order%)
neg.sinh.gamma#=(1!/gamma#-gamma#)/2!
cosh.gamma#=(1!/gamma#+gamma#)/2!
'
FOR i%=1 TO order%\2
sigma#=neg.sinh.gamma#*SIN((2*i%-1)*pi#/(2*order%))
omega#=cosh.gamma#*COS((2*i%-1)*pi#/(2*order%))
b#=2!*ABS(sigma#)
c#=sigma#*sigma#+omega#*omega#
PRINT USING "    b[#] = #.######        c[#] = #.###### ";i%;b#;i%;c#
NEXT i%
'
'  The following section computes the coefficient for the first-order
'  stage present in odd-order filters. Skip over it if order is even.
'
IF (order% MOD 2)=0 THEN GOTO next.order
i%=order%\2+1
sigma#=neg.sinh.gamma#*SIN((2*i%-1)*pi#/(2*order%))
omega#=cosh.gamma#*COS((2*i%-1)*pi#/(2*order%))
c#=ABS(sigma#)
PRINT USING "                        c[#] = #.###### ";i%;c#
'
next.order:
PRINT " "
NEXT order%
'
STOP
```

Listing 5B-1. BASIC program for computing Chebyshev magnitude response.

```
OPTION BASE 0
DIM yval(384)
'
CLS
INPUT"order of filter";order%
INPUT "Passband ripple in dB";ripple#
try.again:
INPUT " passband or stopband? (P/S)",band$
INPUT "Set band edge at ripple limit or 3 dB point? (R/3)",ntype$
ntype$=UCASE$(ntype$)
'
IF UCASE$(band$)<>"P" AND UCASE$(band$)<>"S" GOTO try.again
IF UCASE$(band$)="P" THEN CALL passband.setup
IF UCASE$(band$)="S" THEN CALL stopband.setup
'
CLS
CALL one.cyc.semi.log.box(ylab.val,ylab.incr)
'
'*****************************************************
' deltax% determines the horizontal spacing of computed plot points
' set to 1 for smoothest plots, set to higher values for faster plots
deltax%=1
'
CALL MOVETO(5,13)
PRINT "Chebyshev ";band$;" response for order =";order%;"  ripple =";ripple#
'
IF band$="stopband" THEN GOTO stopband.proc
CALL chebyshev.mag(yval(),deltax%,order%,1,1,ripple#)
CALL magnitude.plot(yval(),deltax%,ymin,ymax,1)
CALL dynamic.halt
'
stopband.proc:
CALL chebyshev.mag(yval(),deltax%,order%,10,1,ripple#)
CALL magnitude.plot(yval(),deltax%,ymin,ymax,1)
CALL dynamic.halt
'
'*********************************************************
'   SUBROUTINES
'*********************************************************
```

```
'   The following subprograms setup values needed for either passband
'   or stopband plotting.
'
SUB passband.setup STATIC
SHARED band$,ylab.val,ylab.incr,ymax,ymin,ripple#
band$="passband"
ylab.val=.5
ylab.incr=.5
'IF ripple#>=.5 THEN ylab.val=.5
'IF ripple#>=.5 THEN ylab.incr=.5*FIX((ripple#/.5)+.99)
ymax=ylab.val
ymin=ymax-8!*ylab.incr
END SUB
'
SUB stopband.setup STATIC
SHARED band$,ylab.val,ylab.incr,ymax,ymin
band$="stopband"
ylab.val=0
ylab.incr=10
ymax=ylab.val
ymin=ymax-8!*ylab.incr
END SUB
'
'**************************************************
'   The following set of subprograms plots a semi-logarithmic
'   grid, one cycle by eight divisions.
'
SUB one.cyc.semi.log.box(ylab.val,ylab.incr) STATIC
CALL box.384.by.224
CALL vertical.grid.line(166,7)
CALL vertical.grid.line(233,7)
CALL vertical.grid.line(281,7)
CALL vertical.grid.line(318,7)
CALL vertical.grid.line(349,7)
CALL vertical.grid.line(375,7)
CALL vertical.grid.line(397,7)
CALL vertical.grid.line(416,7)
FOR iy%=48 TO 216 STEP 28
CALL horizontal.grid.line(iy%,7)
```

```
CALL MOVETO(440,iy%+5)
ylab.val=ylab.val-ylab.incr
PRINT USING "###.#";ylab.val
NEXT iy%
END SUB
'
SUB box.384.by.224 STATIC
LINE (50,20)-(50,244)
LINE (50,20)-(434,20)
LINE(50,244)-(434,244)
LINE(434,20)-(434,244)
END SUB
'
SUB vertical.grid.line(ix%,dot.interval%) STATIC
FOR iy%=20 TO 244 STEP dot.interval%
IF (iy%-20) MOD 28 <>0 THEN PSET(ix%,iy%)
NEXT iy%
END SUB
'
SUB horizontal.grid.line(iy%,dot.interval%) STATIC
FOR ix%=50 TO 434 STEP dot.interval%
PSET (ix%,iy%)
NEXT ix%
END SUB
'
'*******************************************************
'   The following subprogram plots the magnitude response contained in
'   the vector yval( ).
'
SUB magnitude.plot(yval(1),deltax%,ymin,ymax,tracetype%) STATIC
yrange=ymax-ymin
iyold%=244-224*((yval(0)-ymin)/yrange)
iyold2%=INT(244!-224!*((yval(0)-ymin)/yrange))
ixold%=0
FOR ix%=0 TO 383 STEP deltax%
iy%=244-224*((yval(ix%) - ymin)/yrange)
iy2%=INT(244!-224!*((yval(ix%)-ymin)/yrange))
REM LPRINT ix%,yval(ix%),iy%
IF iy%>244 THEN GOTO eof.pb.plot.loop
IF tracetype%=2 THEN PSET(ix%+50,iy%)
```

138

```
IF tracetype%=1 THEN LINE(ixold%+50,iyold%)-(ix%+50,iy%)
IF tracetype%=3 THEN LINE(ixold%+50,iyold2%)-(ix%+50,iy2%)
IF tracetype%=3 THEN LINE(ixold%+50,iyold2%+1)-(ix%+50,iy2%+1)
ixold%=ix%
iyold%=iy%
iyold2%=iy2%
eof.pb.plot.loop:
NEXT ix%
END SUB
'
'*********************************************************
'  The following subprogram computes the Chebyshev magnitude response
'  for a given amount of passband ripple. The number of frequency
'  decades is given by freqcyc%. The order is given by n%.
'
SUB chebyshev.mag(yval(1),deltax%,n%,maxfreq%,freqcyc%,ripple#) STATIC
SHARED ntype$
pi#=3.1415926535898#
m# = 1
epsilon#=SQR(10^(ripple#/10!)-1)
work# = 1!/epsilon#
A = (LOG(work#+SQR(work#*work#-1)))/n%
gamma#=((1+SQR(1+epsilon#*epsilon#))/epsilon#)^(1!/n%)
maxfreqexp%=LOG(maxfreq%)/2.302585093#
rp#=1#
ip#=0#
FOR k%=1 TO n%
x#=(2*k%-1)*pi#/(2!*n%)
i#=.5*(gamma#+1!/gamma#)*COS(x#)
r#=-.5*(gamma#-1!/gamma#)*SIN(x#)
rpt#=ip#*(i#)-rp#*r#
ipt#=rp#*(-i#)-r#*ip#
ip#=ipt#
rp#=rpt#
NEXT k%
H.sub.zero#=SQR(ip#*ip# + rp#*rp#)
IF (n% MOD 2)=0 THEN H.sub.zero#=H.sub.zero#/SQR(1+epsilon#*epsilon#)
FOR ix%=0 TO 383 STEP deltax%
ss#=(10#^(maxfreqexp%+freqcyc%*(ix%-384)/384#))
work#=(EXP(A)+EXP(-A))/2
```

```
IF ntype$="3" THEN ss#=ss#*work#
CALL MOVETO(5,40)
rp#=1#
ip#=0#
FOR k%=1 TO n%
x#=(2*k%-1)*pi#/(2!*n%)
i#=.5*(gamma#+1!/gamma#)*COS(x#)
r#=-.5*(gamma#-1!/gamma#)*SIN(x#)
rpt#=ip#*(i#-ss#)-rp#*r#
ipt#=rp#*(ss#-i#)-r#*ip#
ip#=ipt#
rp#=rpt#
NEXT k%
m#=H.sub.zero#/(SQR(ip#*ip# + rp#*rp#))
yval(ix%)=20*LOG(m#)/2.302585093#
CALL MOVETO(350,13)
PRINT 383-ix%
NEXT ix%
CALL MOVETO(350,13)
PRINT"       "
END SUB
'
'*****************************************************
SUB dynamic.halt STATIC
BEEP
done.loop:
GOTO done.loop
END SUB
```

Listing 5B-2. BASIC program for computing Chebyshev phase response.

```
OPTION BASE 0
DIM yval(384)
'
e# = 2.7182818284592#
m# = 1
pi#=3.1415926535898#
'
CLS
INPUT; "Order of Chebyshev filter";order%
```

```
PRINT " "
INPUT ;"Passband ripple in dB";ripple#
CLS
PRINT USING"Phase response of Chebyshev filters   ripple= #.#";ripple#
CALL two.cyc.semi.log.box
'*******************************************************
' deltax% determines the horizontal spacing of computed plot points
:  set to 1 for smoothest plots, set to higher values for faster plots
deltax%=1
'*******************************************************
'
CALL cheby.phase(yval(),deltax%,order%,10,2,ripple#)
CALL phase.smoother(yval(),deltax%,1)
'
CALL response.plot(yval(),deltax%,-720!,0!,1)
'
CALL dynamic.halt
'***************************************************
'  SUBROUTINES
'***************************************************
'  The following set of subprograms plots a semi-logarithmic grid,
'  two cycles by eight divisions
'
SUB two.cyc.semi.log.box STATIC
CALL box.384.by.224
CALL two.cycle.freq.grid(7)
FOR iy%=48 TO 216 STEP 28
CALL horizontal.grid.line(iy%,7)
NEXT iy%
END SUB
'
SUB box.384.by.224 STATIC
LINE (50,20)-(50,248)
LINE (46,20)-(438,20)
LINE(46,244)-(438,244)
LINE(434,20)-(434,248)
END SUB
'
SUB two.cycle.freq.grid(dot.interval%) STATIC
DIM two.cycle.freq.tics%(17)
```

```
two.cycle.freq.tics%(0)=108
two.cycle.freq.tics%(1)=141
two.cycle.freq.tics%(2)=166
two.cycle.freq.tics%(3)=184
two.cycle.freq.tics%(4)=200
two.cycle.freq.tics%(5)=213
two.cycle.freq.tics%(6)=224
two.cycle.freq.tics%(7)=233
two.cycle.freq.tics%(8)=242
two.cycle.freq.tics%(9)=300
two.cycle.freq.tics%(10)=333
two.cycle.freq.tics%(11)=358
two.cycle.freq.tics%(12)=376
two.cycle.freq.tics%(13)=392
two.cycle.freq.tics%(14)=405
two.cycle.freq.tics%(15)=416
two.cycle.freq.tics%(16)=425
FOR tic.index%=0 TO 16
ix%=two.cycle.freq.tics%(tic.index%)
FOR iy%=20 TO 244 STEP dot.interval%
IF (iy%-20) MOD 28 <>0 THEN PSET(ix%,iy%)
NEXT iy%
LINE(ix%,244)-(ix%,248)
NEXT tic.index%
LINE(242,20)-(242,244)
END SUB
'
SUB horizontal.grid.line(iy%,dot.interval%) STATIC
LINE(46,iy%)-(50,iy%)
FOR ix%=50 TO 434 STEP dot.interval%
PSET (ix%,iy%)
NEXT ix%
LINE(434,iy%)-(438,iy%)
END SUB
'
'*********************************************************
'
'   The following subprogram removes discontinuities in the phase
'   response arising from the use of only the principal value for
'   the arctan function.
'
```

```
SUB phase.smoother(yval(1),deltax%,angletype%) STATIC
full.circle.offset=0
FOR ix%=deltax% TO 383 STEP deltax%
yval(ix%)=yval(ix%)+full.circle.offset
IF ABS(yval(ix%-deltax%)-yval(ix%))<90! THEN GOTO offset.ok
IF yval(ix%-deltax%)>yval(ix%) THEN GOTO increase.offset
yval(ix%)=yval(ix%)-180
full.circle.offset=full.circle.offset-180
GOTO offset.ok
increase.offset:
yval(ix%)=yval(ix%)+180
full.circle.offset=full.circle.offset+180
offset.ok:
NEXT ix%
END SUB
'
'*********************************************************

'   The following subprogram plots the phase response contained in
'   the vector yval().
'
SUB response.plot(yval(1),deltax%,ymin,ymax,tracetype%) STATIC
yrange=ymax-ymin
iyold%=244-224*((yval(0)-ymin)/yrange)
iyold2%=INT(244!-224!*((yval(0)-ymin)/yrange))
ixold%=0
FOR ix%=0 TO 383 STEP deltax%
iy%=244-224*((yval(ix%) - ymin)/yrange)
iy2%=INT(244!-224!*((yval(ix%)-ymin)/yrange))
IF iy%>244 THEN GOTO eof.pb.plot.loop
IF tracetype%=2 THEN PSET(ix%+50,iy%)
IF tracetype%=1 THEN LINE(ixold%+50,iyold%)-(ix%+50,iy%)
IF tracetype%=3 THEN LINE(ixold%+50,iyold2%)-(ix%+50,iy2%)
IF tracetype%=3 THEN LINE(ixold%+50,iyold2%+1)-(ix%+50,iy2%+1)
ixold%=ix%
iyold%=iy%
iyold2%=iy2%
eof.pb.plot.loop:
NEXT ix%
END SUB
'
```

```
'*******************************************************
'   The following subprogram computes the Chebyshev phase response
'   for a given amount of passband ripple. The number of frequency
'   decades is given by freqcyc%. The order is given by n%.
'
SUB cheby.phase(yval(1),deltax%,n%,maxfreq%,freqcyc%,ripple#) STATIC
SHARED ntype$
pi#=3.1415926535898#
m# = 1
epsilon#=SQR(10^(ripple#/10!)-1)
work# = 1!/epsilon#
A = (LOG(work#+SQR(work#*work#-1)))/n%
gamma#=((1+SQR(1+epsilon#*epsilon#))/epsilon#)^(1!/n%)
maxfreqexp%=LOG(maxfreq%)/2.302585093#
rp#=1#
ip#=0#
FOR k%=1 TO n%
x#=(2*k%-1)*pi#/(2!*n%)
i#=.5*(gamma#+1!/gamma#)*COS(x#)
r#=-.5*(gamma#-1!/gamma#)*SIN(x#)
rpt#=ip#*(i#)-rp#*r#
ipt#=rp#*(-i#)-r#*ip#
ip#=ipt#
rp#=rpt#
NEXT k%
H.sub.zero#=SQR(ip#*ip# + rp#*rp#)
IF (n% MOD 2)=0 THEN H.sub.zero#=H.sub.zero#/SQR(1+epsilon#*epsilon#)
FOR ix%=0 TO 383 STEP deltax%
ss#=(10^(maxfreqexp%+freqcyc%*(ix%-384)/384#))
work#=(EXP(A)+EXP(-A))/2
IF ntype$="3" THEN ss#=ss#*work#
CALL MOVETO(5,40)
rp#=1#
ip#=0#
FOR k%=1 TO n%
x#=(2*k%-1)*pi#/(2!*n%)
i#=.5*(gamma#+1!/gamma#)*COS(x#)
r#=-.5*(gamma#-1!/gamma#)*SIN(x#)
rpt#=ip#*(i#-ss#)-rp#*r#
ipt#=rp#*(ss#-i#)-r#*ip#
```

```
ip#=ipt#
rp#=rpt#
NEXT k%
p#=-ATN(ip#/rp#)
yval(ix%)=p#*180#/pi#
CALL MOVETO(350,13)
PRINT 383-ix%
NEXT ix%
CALL MOVETO(350,13)
PRINT"        "
END SUB
'
'********************************************************
'
SUB dynamic.halt STATIC
BEEP
done.loop:
GOTO done.loop
END SUB
```

Listing 5C-1. BASIC program for computing Chebyshev transient response.

```
OPTION BASE 0
DIM yval(384)
'
deltax%=1
pi#=3.1415926535898#
'
INPUT"order of Chebyshev filter ";n%
INPUT"Impulse or step response ? (I/S) ";trans.type$
INPUT "Passband ripple in dB ";ripple#
IF UCASE$(trans.type$)<>"I" AND UCASE$(trans.type$)<>"S" GOTO try.again
IF UCASE$(trans.type$)="S" THEN GOTO step.job
'
impulse.job:
CLS
PRINT USING"Lowpass Chebyshev impulse response.  order=##";n%;
PRINT USING"  ripple = #.## dB";ripple#
CALL impulse.box
```

```
CALL chebyshev.impulse(yval(),n%,deltax%,pi#,ripple#)
CALL impulse.plot(yval(),-.2,.7)
CALL dynamic.halt
'
step.job:
CLS
PRINT USING"Lowpass Chebyshev step response.   order=##";n%
CALL impulse.box
CALL chebyshev.impulse(yval(),n%,deltax%,pi#,ripple#)
CALL integrate(yval(),yval.scaler)
CALL step.plot(yval(),0!,1.4,yval.scaler)
CALL dynamic.halt
'
'***********************************************************
'   SUBROUTINES
'***********************************************************
'
SUB dynamic.halt STATIC
BEEP
done.loop:
GOTO done.loop
END SUB
'

'***********************************************************
'
'   The following set of subprograms plots the grid for the impulse response.
'
SUB impulse.box STATIC
CALL box.384.by.224
CALL imp.grid
END SUB
'
SUB box.384.by.224 STATIC
LINE(50,20)-(50,244)
LINE(50,20)-(434,20)
LINE(50,244)-(434,244)
LINE(434,20)-(434,244)
END SUB
'
SUB imp.grid STATIC
LINE(50,212)-(434,212)
```

146

```
LINE(50,180)-(434,180)
LINE(50,148)-(434,148)
LINE(50,116)-(434,116)
LINE(50,84)-(434,84)
LINE(50,52)-(434,52)
LINE(150,20)-(150,244)
LINE(250,20)-(250,244)
LINE(350,20)-(350,244)
END SUB
'**************************************************
'
'   The following subprogram plots the impulse response data
'   contained in the vector yval().
'
SUB impulse.plot(yval(1),ymin,yrange) STATIC
iyold%=244-224*((yval(0)-ymin)/yrange)
ixold%=0
'
FOR ix%=1 TO 383
iy%=244-224*((yval(ix%)-ymin)/yrange)
IF iy%>244 THEN GOTO eof.plot.loop
LINE(ixold%+50,iyold%)-(ix%+50,iy%)
ixold%=ix%
iyold%=iy%
eof.plot.loop:
NEXT ix%
'
END SUB
'
'**************************************************
'   The following subprogram integrates the impulse response to
'   obtain the step response.
'
SUB integrate(yval(1),yval.scaler) STATIC
'
FOR ix2%=1 TO 383
yval(ix2%)=yval(ix2%-1)+yval(ix2%)
NEXT ix2%
IF n%<4 THEN yval.scaler=yval(382)
END SUB
```

```
'*********************************************************
'   The following subprogram plots the step response data contained
'   in the vector yval().
'
SUB step.plot(yval(1),ymin,yrange,yval.scaler) STATIC
'
iyold%=244-224*((yval(0)/yval.scaler-ymin)/yrange)
ixold%=0
FOR iix%=1 TO 383
iy%=244-224*((yval(iix%)/yval.scaler)/yrange)
IF iy%>244 THEN GOTO eof.plot.loop
REM PSET(iix%+50,iy%)
LINE(ixold%+50,iyold%)-(iix%+50,iy%)
ixold%=iix%
iyold%=iy%
eof.plot.loop:
NEXT iix%
END SUB
'

'*********************************************************
'   The following subprogram computes the impulse response of
'   a Chebyshev filter of order n%.
'
SUB chebyshev.impulse(yval(1),n%,deltax%,pi#,ripple#) STATIC
FOR ix%=0 TO 383 STEP deltax%
CALL MOVETO(420,13)
PRINT 383-ix%
h.of.t#=0!
t=.05*ix%
epsilon#=SQR(10^(ripple#/10!)-1)
work#=1!/epsilon#
A=(LOG(work#+SQR(work#*work#-1)))/n%
norm.factor#=(EXP(A)+EXP(-A))/2!
gamma#=((1+SQR(1+epsilon#*epsilon#))/epsilon#)^(1!/n%)
'
'   Compute H.zero#
'
rp#=1
ip#=0
FOR k%=1 TO n%
x#=(2*k%-1)*pi#/(2!*n%)
```

148

```
i#=.5*(gamma#+1!/gamma#)*COS(x#)/norm.factor#
r#=-.5*(gamma#-1!/gamma#)*SIN(x#)/norm.factor#
rpt#=ip#*(i#)-rp#*r#
ipt#=rp#*(-i#)-r#*ip#
ip#=ipt#
rp#=rpt#
NEXT k%
H.zero#=SQR(ip#*ip#+rp#*rp#)
IF(n% MOD 2)=0 THEN H.zero#=H.zero#/SQR(1+epsilon#*epsilon#)
FOR r%=1 TO n%\2
.
x#=(2*r%-1)*pi#/(2!*n%)
sigma=-.5*(gamma#-1!/gamma#)*SIN(x#)/norm.factor#
omega=.5*(gamma#+1!/gamma#)*COS(x#)/norm.factor#
.
' Compute Lr and Mr
.
L#=1
M#=0
.
FOR ii%=1 TO n%
IF ii%=r% GOTO end.I.loop
.
x#=(2*ii%-1)*pi#/(2!*n%)
R#= sigma - (-.5*(gamma#-1!/gamma#))*SIN(x#)/norm.factor#
I#= omega - (.5*(gamma#+1!/gamma#))*COS(x#)/norm.factor#
.
LT#=L# * R# - M# * I#
MT#= L# * I# + R# * M#
L# = LT#
M# = MT#
end.I.loop:
NEXT ii%
.
L# = LT#  / (LT# * LT# + MT# *MT#)
M# = -MT#  / (LT# *LT# + MT# * MT#)
.
cos.part# = 2*L#*EXP(sigma*t)*COS(omega*t)
sin.part# = 2*M#*EXP(sigma*t)*SIN(omega*t)
.
h.of.t# = h.of.t# + cos.part# - sin.part#
NEXT r%
```

```
IF (n% MOD 2) = 0 THEN GOTO end.T.loop

'*********************************************************
' The following section computes the real exponential component
' present in odd-order responses.
K#=1
L#=1
M#=0
r%=(n%+1)/2

x#=(2*r%-1)*pi#/(2!*n%)
sigma=-.5*(gamma#-1!/gamma#)*SIN(x#)/norm.factor#
omega=.5*(gamma#+1!/gamma#)*COS(x#)/norm.factor#
'
FOR iii%=1 TO n%
IF iii%=r% GOTO end.III.loop
'
x#=(2*iii%-1)*pi#/(2!*n%)
R#= sigma - (-.5*(gamma#-1!/gamma#))*SIN(x#)/norm.factor#
I#= omega - (.5*(gamma#+1!/gamma#))*COS(x#)/norm.factor#
'
LT#=L# * R# - M# * I#
MT#= L# * I# + R# * M#
L# = LT#
M# = MT#

end.III.loop:
NEXT iii%
K# = LT# / (LT# * LT# + MT# *MT#)
h.of.t# = h.of.t# + K#*EXP(sigma*t)
'
end.T.loop:
yval(ix%)=h.of.t# * H.zero#
'
IF yval(ix%)>ymax THEN ymax=yval(ix%)
IF yval(ix%)<ymin THEN ymin=yval(ix%)
NEXT ix%
CALL MOVETO(420,13)
PRINT "    "
END SUB
```

Chapter 6

Active Filters

A CTIVE FILTERS ARE ANALOG FILTER CIRCUITS THAT IN- clude active elements such as transistors or op amps to obtain performance improvements or economic advantages over corresponding passive filter designs. There has been a great deal of work done in the area of active filter design as indicated by the numerous books and articles that have been written on various facets of the subject. In this book we will limit ourselves to only the most popular active filter types and some simple procedures for their design. Knowledge of all the subtleties, derivations, and specialized configurations is not necessary to be able to use active filters in many moderately demanding signal processing applications. If such information is needed, it can be obtained elsewhere.

At present, the most straightforward approach for the design of active filters involves designing a number of first- and second-order filter stages and then cascading them to form a complete filter of the desired order as shown in Fig. 6-1. The following sections present design procedures for first- and second-order stages, along with some guidelines for cascading stages to realize higher-order filters.

6.1 LOWPASS FILTER CIRCUITS

Transfer functions of an n-th order all-pole lowpass filter are

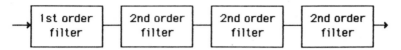

Fig. 6-1. Higher-order filter realized as a cascade of first- and second-order stages.

shown in Fig. 6-2. The general form given as Equation 6.1-1 can be factored into a product of normalized second-order transfer functions as in Equation 6.1-2 when n is even, or into a product of normalized second-order functions with one first-order function as in Equation 6.1-3 when n is odd. The particular values chosen for the coefficients b_i and c_i will determine the shape of the filter's response. Butterworth and Chebyshev filters both have transfer functions of this form, and appropriate coefficient values for them can be found in Chapter 4 or 5.

Design Procedure 6-1. To design lowpass active filters, perform the following steps:

$$H(s) = \frac{K a_0}{a_0 + a_1 s + a_2 s^2 + \ldots + a_{n-1} s^{n-1} + s^n} \qquad \text{(Eq. 6.1-1)}$$

$$H(s) = \prod_{i=1}^{n/2} \left(\frac{(2\pi f_c)^2 K_i c_i}{s^2 + 2\pi f_c b_i s + (2\pi f_c)^2 c_i} \right) \qquad \text{(Eq. 6.1-2)}$$

$$H(s) = \frac{2\pi f_c K_0 c_0}{s + 2\pi f_c c_0} \prod_{i=1}^{(n-1)/2} \left(\frac{(2\pi f_c)^2 K_i c_i}{s^2 + 2\pi f_c b_i s + (2\pi f_c)^2 c_i} \right)$$

$$\text{(Eq. 6.1-2)}$$

Fig. 6-2. Transfer function of an all-pole lowpass filter.

152

1. Select the overall response shape (Butterworth, Chebyshev, etc.) needed to satisfy design requirements.

2. Using data from Chapter 4 or 5 as appropriate, determine the minimum filter order needed to satisfy design requirements.

3. If the filter order n determined in Step 2 is even, design n/2 second-order lowpass stages using Design Procedure 6-3.

4. If the filter order n determined in Step 2 is odd, design (n – 1)/2 second-order lowpass stages using Design Procedure 6-3 and one first-order stage using Design Procedure 6-2.

5. Cascade the stages designed in Step 3 or 4 to form the complete filter. When sections are to be cascaded, it is important for each stage to have a high input impedance and reasonably low output impedance to prevent excessive loading of any stage and the consequent distortion of the response that this could cause.

6.1.1 First-Order Lowpass Stages. Figure 6-3 shows active filter circuits that can be used to implement the first-order stage of a lowpass filter. Circuit A is used for stages with gain values greater than one, and circuit B is used for stages with unity gain.

Design Procedure 6-2. To design a first-order lowpass active filter stage perform the following steps:

1. Using formulas, tables or programs from Chapter 4 or 5, determine the appropriate value of the coefficient "c" for the normalized transfer function of the desired first-order lowpass stage.

2. Determine the desired stage gain K, and cutoff frequency f_c.

3. Pick a readily available standard value for C_1 which is close to $10^{-5}/f_c$.

4. Compute the resistor values using the formulas in Fig. 6-4.

5. Select an op amp that has an input impedance Z_{in} and an open loop gain A_{vo} meeting the conditions given in Fig. 6-4.

6. If necessary, tune the cutoff frequency f_c by adjusting the value of R_1. Increase the stage gain K by increasing the ratio of R_3 to R_2, or conversely, decrease the gain by decreasing this ratio.

6.1.2 Second-Order All Pole Lowpass Stages. Several different circuits can be used to implement second-order stages of Butterworth or Chebyshev lowpass filters. Figure 6-5 shows a *Salen-Key* or *voltage-controlled voltage source* (VCVS) filter circuit and Fig. 6-7 shows an *infinite gain multiple feedback* (MFB) lowpass filter circuit—either of which can be used for stages having $Q < 10$

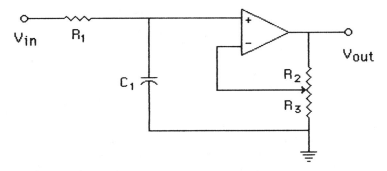

circuit for first-order stage with K>1

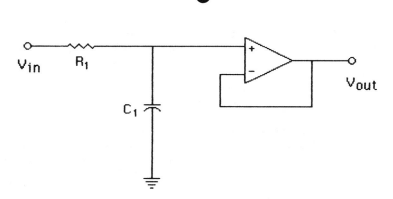

circuit for first-order stage with K=1

B

Fig. 6-3. First-order lowpass active filter circuits.

and $KQ \leq 100$. Figure 6-9 shows a *biquadratic* (biquad) lowpass filter circuit that can be used for stages in which $Q \leq 100$. (Remember that for lowpass stages $Q = \sqrt{c}/b$.)

Design Procedure 6-3. To design an all-pole second-order lowpass filter stage, perform the following steps:

1. Select the circuit configuration to be used. The VCVS offers moderate performance and noninverting gain while using just

passive element values:

$$C_1 \cong \frac{10^{-5}}{f_c}$$

$$R_2 = KR_1$$

$$R_1 = \frac{1}{2\pi c f_c C_1}$$

$$R_3 = \frac{KR_1}{K-1}$$

op amp parameters:

$$A_{vo} \geq 50\sqrt{\frac{K^2 c^2}{c^2+1}}$$

$$Z_{in} \geq 10 R_1$$

Fig. 6-4. Design formulas for first-order lowpass active filter stages.

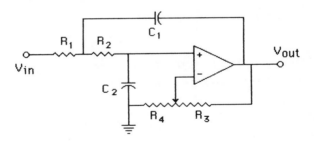

Tuning Data:

adjust K by varying $\dfrac{R_3}{R_4}$ (increasing $\dfrac{R_3}{R_4}$ increases K)

tune f_c by varying both R_1 and R_2 by equal percentages

(increasing R_1 and R_2 decreases f_c)

Fig. 6-5. Second-order VCVS lowpass active filter circuit.

passive element values:

$$C_1 \cong \frac{10^{-5}}{f_c} \qquad C_2 \leq \frac{C_1[b^2 + 4c(K-1)]}{4c}$$

$$R_1 = \frac{1}{\pi f_c \left[bC_1 + \sqrt{C_1(b^2C_1 + 4cC_1(K-1) - 4cC_2)}\right]}$$

$$R_2 = \frac{1}{(2\pi f_c)^2 R_1 c C_1 C_2}$$

$$R_3 = K(R_1 + R_2) \quad ; K > 1 \qquad\qquad R_4 = \frac{K(R_1 + R_2)}{K-1} \quad ; K > 1$$
$$= \text{short circuit} \; ; K = 1 \qquad\qquad\qquad = \text{open circuit} \quad ; K = 1$$

Pick standard values of R and R close to the calculated values and which satisfy

$$\frac{R_3}{R_4} = K - 1$$

op amp parameters:

$$A_{vo} \geq 50 \sqrt{\frac{K^2 c^2}{(c-1)^2 + b^2}} \qquad\qquad Z_{in} \geq 10(R_1 + R_2)$$

Fig. 6-6. Design formulas for second-order VCVS lowpass filter stages.

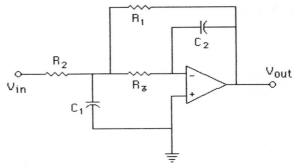

Tuning Data:

Adjust gain K by varying R_1 (increasing R_1 increases K)

Tune f_c by varying R_3 (increasing R_3 decreases f_c)

Tune f_m by varying R_2 (increasing R_2 increases f_m)

Fig. 6-7. Second-order MFB lowpass active filter circuit.

one op amp per stage. The MFB circuit also uses one op amp per stage and offers moderate performance with inverting gain. The biquad circuit offers superior performance and a choice of either inverting or noninverting gain, but requires three op amps per stage.

2. Using formulas, tables, or programs from Chapter 4 or 5, determine the appropriate value of the coefficients b and c for the normalized transfer function of the desired stage gain K and cut-off frequency f_c.

3. Pick a readily available standard value for C_1 which is close to $10^{-5}/f_c$.

4. Determine the remaining capacitor values (if any), and compute the resistor values using the design formulas in Fig. 6-6 (VCVS), Fig. 6-8 (MFB), or Fig. 6-10 (biquad) as appropriate.

5. Select op amps which have input impedance Z_{in} and open loop gain A_{vo} meeting the conditions given in the appropriate set of design formulas.

6. If necessary, tune the cutoff frequency f_c, peaking frequency f_m, and stage gain K by adjusting the resistor values as shown in each schematic. These adjustments interact with each other and thus may have to be repeated several times to achieve the desired performance.

passive element values:

$$C_1 \cong \frac{10^{-5}}{f_c} \qquad\qquad C_2 \leq \frac{b^2 C_1}{4c(K+1)}$$

$$R_1 = \frac{K+1}{\pi f_c \left[bC_1 + \sqrt{C_1(b^2 C_1 - 4cC_2(K+1))} \right]}$$

$$R_2 = \frac{R_1}{K} \qquad\qquad R_3 = \frac{1}{2\pi c\, f_c R_1 C_1 C_2}$$

op amp parameters:

$$A_{vo} \geq 50 \sqrt{\frac{K^2 c^2}{(c-1)^2 + b^2}}$$

$$Z_{in} \geq 10 \left[\frac{R_1 R_2}{R_1 + R_2} + R_3 \right]$$

Fig. 6-8. Design formulas for second-order MFB lowpass filter stages.

Example 6-1. Using two second-order VCVS stages, design a fourth-order lowpass Butterworth filter having a cutoff frequency of 3 kHz and an overall gain of 12. Use standard capacitor values and standard 1% resistor values.

Solution. From Table 4-1 determine that for the lowpass Butterworth case of $n = 4$,

$$b_1 = 0.75367 \qquad c_1 = 1.00$$
$$b_2 = 1.841759 \qquad c_2 = 1.00$$

The overall gain K can be arbitrarily factored into two separate stage gain values of $K_1 = 4$ and $K_2 = 3$. Using these coefficients and gain values in conjunction with Design Procedure 6-3 and formulas from Fig. 6-6, you can easily determine the element values needed for each second-order stage.

For the first stage:

$$C_1 \cong 3.333 \times 10^{-8} \text{ F}$$

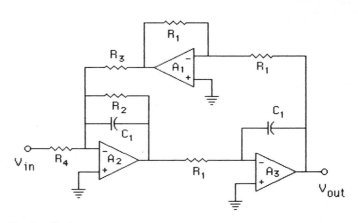

Tuning Data:

adjust K by varying R_4 (increasing R_4 decreases K)

tune f_C by varying R_3 (increasing R_3 decreases f_C)

Fig. 6-9. Second-order biquad lowpass active filter circuit.

passive element values:

$$C_1 \cong \frac{10^{-5}}{f_C}$$

$$R_1 = \frac{1}{2\pi f_C C_1} \qquad\qquad R_3 = \frac{R_1}{c}$$

$$R_2 = \frac{R_1}{b} \qquad\qquad\qquad R_4 = \frac{R_1}{cK}$$

op-amp parameters:

$$A_{vo} \geq 50 \sqrt{\frac{K^2 c^2}{(c-1)^2 + b^2}}$$

$$Z_{in} \geq 10R_1 \qquad \text{for amplifiers } A_1 \text{ and } A_3$$

$$Z_{in} \geq 10R_4 \qquad \text{for amplifier } A_2$$

Fig. 6-10. Design formulas for second-order biquad lowpass filter stages.

Polystyrene capacitors are readily available in a standard value of 33000 pF. Therefore let C_1 = 33000 pF = 3.3×10^{-8} F. Then we compute

$$C_2 \leq 1.038 \times 10^{-7} \text{ F}$$

Polycarbonate or mylar capacitors are readily available in a standard value of 0.1 μF. Therefore let C_2 = 0.1 μF = 1×10^{-7} F. You then use these values of C_1 and C_2 to compute

$$R_1 = 2222 \text{ ohms}$$

The closest standard value for 1% resistors is 383 KΩ. You then use this value to compute

$$R_2 = 383.8 \text{ ohms}$$

The closest standard value of 1% resistors is 383 KΩ. You then use this value to compute

$$R_3 \cong 10.423 \text{ K}\Omega$$
$$R_4 \cong \Omega\ 3.474 \text{ K}\Omega$$
$$R_3/R_4 = 3$$

From the table of standard 1% resistor values, you can determine that the best choices are 10.2 KΩ for R_3 and 3.40 KΩ for R_4, which results in R_3/R_4 = 3. In a similar fashion, you can determine the second stage values:

$$
\begin{aligned}
C_1 &= 33000 \text{ pF} \\
C_2 &= 82000 \text{ pF} \\
R_1 &= 1.05 \text{ K}\Omega \\
R_2 &= 976 \ \Omega \\
R_3 &= 6.19 \text{ K}\Omega \\
R_4 &= 3.09 \text{ K}\Omega
\end{aligned}
$$

The complete fourth-order filter is shown in Fig. 6-11.

6.2 HIGHPASS FILTER CIRCUITS

If you apply the lowpass-to-highpass transformation of Section

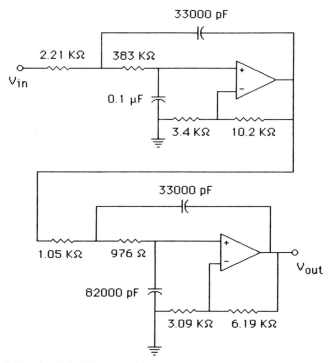

Fig. 6-11. Circuit for Example 6-1.

3.3 to the lowpass stage Equations 6.1-2 and 6.1-3, you obtain the factored highpass transfer functions shown in Fig. 6-12. Equation 6.2-1 is for even orders and Equation 6.2-2 is for odd orders. The values of b_i and c_i are the exact same values used for the corresponding Butterworth or Chebyshev lowpass filter.

Design Procedure 6-4. To design highpass active filters, perform the following steps:

1. Select the overall response shape and determine the filter order needed to satisfy design requirements.

2. If the filter order n determined in Step 1 is even, design n/2 second-order highpass stages using Design Procedure 6-6.

3. If the filter order n determined in Step 1 is odd, design $(n - 1)/2$ second-order highpass stages using Design Procedure 6-6 and one first-order highpass stage using Design Procedure 6-5.

4. Cascade the stages designed in Step 2 or 3 to form the complete filter.

162

6.2-1 First-Order Highpass Stages. Figure 6-13 shows active filter circuits that can be used to implement a first-order stage of a highpass filter. Circuit A is used for stages with gain values greater than one, and circuit B is used for stages with unity gain.

Design Procedure 6-5. To design a first-order highpass active filter stage, perform the following steps:

1. Using formulas, tables, or programs from Chapter 4 or 5, determine the appropriate value of the coefficient c for the transfer function of the desired first-order highpass stage.

2. Determine the desired stage gain K, and cutoff frequency f_c.

3. Pick a readily available standard value for C_1 that is close to $10^{-5}/f_c$.

4. Compute the resistor values using the design formulas in Fig. 6-14.

5. Select an op amp that has an input impedance Z_{in} and an open loop gain A_{vo} that meet both conditions given in Fig. 6-14.

6. If necessary, tune the cutoff frequency f_c by adjusting the value of R_1. Increase the stage gain K by increasing the ratio of R_3 to R_2, or conversely, decrease the gain by decreasing this ratio.

6.2.2 Second-Order Highpass Stages. Several different circuits can be used to implement second-order stages of Butterworth or Chebyshev highpass filters. Figure 6-15 shows a VCVS highpass filter circuit and Fig. 6-17 shows an MFB highpass filter circuit. Either of these circuits can be used for stages having $Q \leq 10$ and $KQ \leq 100$. Figure 6-19 shows a biquad highpass filter circuit that can be used for stages in which $Q \leq 100$.

$$H(s) = \prod_{i=1}^{n/2} \left(\frac{c_i K_i s^2}{c_i s^2 + 2\pi b_i f_c s + (2\pi f_c)^2} \right) \qquad \text{(Eq. 6.2-1)}$$

$$H(s) = \frac{c_0 K_0 s}{c_0 s + \omega_c} \prod_{i=1}^{(n-1)/2} \left(\frac{c_i K_i s^2}{c_i s^2 + 2\pi b_i f_c s + (2\pi f_c)^2} \right)$$

$$\text{(Eq. 6.2-2)}$$

Fig. 6-12. Factored highpass transfer functions.

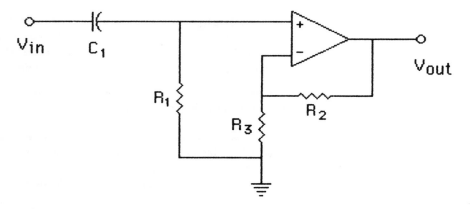

circuit for first-order stage with K>1

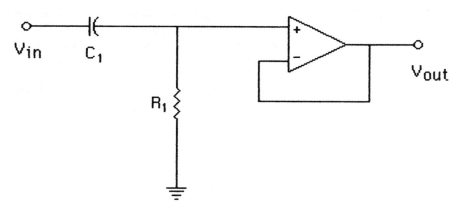

circuit for first-order stage with K=1

Fig. 6-13. First-order highpass active filter circuits.

passive element values:

$$C_1 \cong \frac{10^{-5}}{f_c}$$

$$R_2 = K R_1$$

$$R_1 = \frac{c}{2\pi f_c C_1}$$

$$R_3 = \frac{K R_1}{K-1}$$

op amp parameters:

$$A_{vo} \geq 50 \sqrt{\frac{K^2 c^2}{c^2 + 1}}$$

$$Z_{in} \geq 10 R_1$$

Fig. 6-14. Design formulas for first-order highpass active filter stages.

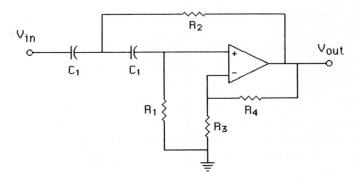

Tuning Data:

adjust K by varying $\frac{R_4}{R_3}$ (increasing $\frac{R_4}{R_3}$ increases K)

tune f_c by varying both R_1 and R_2 by equal percentages

(increasing R_1 and R_2 decreases f_c)

Fig. 6-15. Second-order VCVS highpass filter circuit.

165

$$R_1 = \frac{1}{\pi f_c C_1 \left[b + \sqrt{b^2 + 8c(K-1)} \right]}$$

$$R_2 = \frac{c}{(2\pi f_c)^2 R_1 C_1^2}$$

$$R_3 = \frac{K R_1}{K-1} \qquad ; K > 1$$

$$= \text{open circuit} \qquad ; K = 1$$

$$R_4 = K R_1 \qquad ; K > 1$$

$$= \text{short circuit} \qquad ; K = 1$$

The input impedance of the op-amp used should be $10 R_2$ or greater.

This circuit should be used only for stages in which $Q \leq 10$ and $KQ \leq 100$.

Fig. 6-16. Design formulas for second-order VCVS highpass filter stages.

Design Procedure 6-6. To design a second-order stage of a Butterworth or Chebyshev highpass filter, perform the following steps:

1. Using formulas, tables, or programs from Chapter 4 or 5, determine the appropriate value of the coefficients b and c for the transfer function of the desired second-order highpass stage.

2. Determine the desired stage gain K and cutoff frequency f_c.

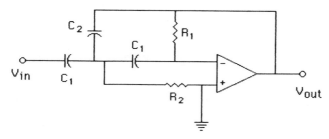

Fig. 6-17. Second-order MFB highpass filter circuit.

Compute the value of $C_2 = C_1/K$. Ideally, a value of K should be selected that will yield readily available standard values for both C_1 and C_2.

$$R_1 = \frac{c(2C_1 + C_2)}{2\pi f_c b C_1 C_2} \qquad R_2 = \frac{b}{2\pi f_c(2C_1 + C_2)}$$

The input impedance of the op amp used should be $10R_1$ or greater.

This circuit should be used only for stages in which $Q \leq 10$ and $KQ \leq 100$.

Fig. 6-18. Design formulas for second-order MFB highpass filter stages.

3. Pick a readily available standard value for C_1 that is close to $10^{-5}/f_c$.

4. Determine the remaining capacitor values (if any), and compute the resistor values using the design formulas in Fig. 6-16 (VCVS), Fig. 6-18 (MFB), or Fig. 6-20 (biquad) as appropriate.

5. Select an op amp that has an input impedance and an open loop gain meeting the conditions shown in the appropriate set of design formulas.

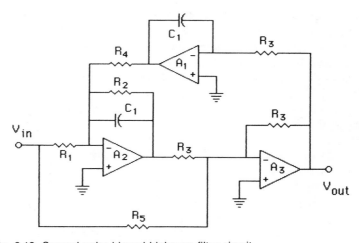

Fig. 6-19. Second-order biquad highpass filter circuit.

$$R_1 = \frac{c}{2\pi f_c b K C_1} \qquad R_2 = K R_1 \qquad R_3 = \frac{1}{2\pi f_c C_1}$$

$$R_4 = \frac{c}{(2\pi f_c)^2 C_1^2 R_3} \qquad R_5 = \frac{R_3}{K}$$

The input impedance of op amps A_1 and A_3 should be $10 R_3$ or greater, and the input impedance of A_2 should be $10 R_1$ or greater.

This circuit should be used only for stages in which $Q \leq 100$.

Fig. 6-20. Design formulas for second-order biquad highpass filter stages.

6. If necessary, tune the cutoff frequency f_c, peaking frequency f_m, and stage gain K by adjusting the resistor values as shown in each schematic. These adjustments can interact with each other and thus may have to be repeated several times to achieve the desired performance.

6.3 BANDPASS FILTER CIRCUITS

As discussed in Section 3.4, bandpass filters can be realized in two different ways. If the bandwidth of the deisred filter is relatively wide you can simply cascade a lowpass filter and a highpass filter to obtain a bandpass response. However, for relatively narrowband bandpass filters you must use the lowpass-to-bandpass transformation discussed in Section 3.4.2. This transformation will produce a second-order bandpass stage for each first-order lowpass stage, and a pair of second-order bandpass stages for each single second-order lowpass stage. The design procedures will vary slightly for these two cases.

As with the lowpass and highpass cases, there are several different circuits that can be used to implement stages of Butterworth or Chebyshev narrowband bandpass filters. Figure 6-21 shows a VCVS bandpass filter circuit and Fig. 6-22 shows an MFB bandpass circuit. Either of these circuits can be used for stages having $Q \leq 10$ and $KQ \leq 100$. Figure 6-23 shows a biquad bandpass filter circuit that can be used for stages in which $Q \leq 100$.

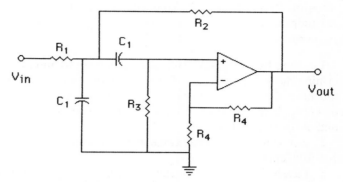

Fig. 6-21. Second-order VCVS bandpass filter circuit.

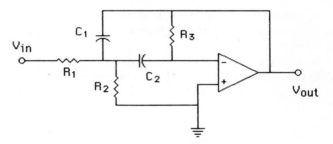

Fig. 6-22. Second-order MFB bandpass filter circuit.

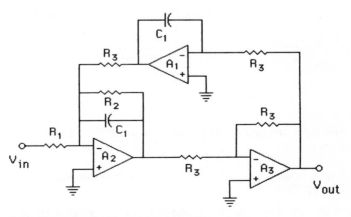

Fig. 6-23. Second-order biquad bandpass filter circuit.

Design Procedure 6-7. To design a second-order narrow-band bandpass filter stage corresponding to a first-order lowpass stage, perform the following steps:

1. Using formulas, tables or programs from Chapter 4 or 5, determine the value of the coefficient c for the transfer function of the normalized lowpass stage corresponding to the desired second-order bandpass stage. If the entire filter consists of just this second-order stage, then c = 1.
2. Determine the desired stage gain K, quality factor Q, and center frequency f_o.
3. Pick a readily available standard value for C_1 that is close to $10^{-5}/f_o$.
4. Determine the remaining capacitor values (if any), and compute the resistor values using the design formulas in Fig. 6-24 (VCVS), Fig. 6-25 (MFB), or Fig. 6-26 (biquad) as appropriate.
5. Select an op amp that has an input Z_{in} and an open loop gain A_{vo} meeting the conditions given in the appropriate set of design formulas.
6. Construct the circuit in accordance with Fig. 6-21, 6-22, or 6-23 using the component values determined in Steps 4 and 5.
7. If necessary, tune the center frequency f_o, quality factor

$$R_1 = \frac{Q}{\pi f_0 \, K c \, C_1}$$

$$R_2 = \frac{-Q}{\pi f_0 \, C_1 \left[c - \sqrt{((K-1)c)^2 + 8Q^2} \right]}$$

$$R_3 = \frac{R_1 + R_2}{(2\pi f_0)^2 \, C_1^2 \, R_1 R_2}$$

$$R_4 = 2R_3$$

The input impedance of the op amp used should be $10R_3$ or greater.

This circuit should be used only for stages in which $Q \leq 10$ and $KQ \leq 100$.

Fig. 6-24. Design formulas for second-order VCVS bandpass filter stages corresponding to first-order lowpass stages.

Pick the smallest readily available standard value for C_2 such that

$$C_2 > \frac{K c^2}{Q^2} C_1$$

$$R_1 = \frac{Q}{2\pi f_0 K c\, C_1}$$

$$R_2 = \frac{cQ}{2\pi f_0 [\, Q^2 (C_1 + C_2) - K c^2 C_1\,]}$$

$$R_3 = \frac{Q(C_1 + C_2)}{2\pi f_0\, c\, C_1 C_2}$$

The input impedance of the op amp used should be $10 R_3$ or greater.

This circuit should be used only for stages in which $Q \le 10$ and $KQ \le 100$.

Fig. 6-25. Design formulas for second-order MFB bandpass filter stages corresponding to first-order lowpass stages.

Q, and stage gain K by adjusting the resistor values as indicated in each schematic. These adjustments can interact with each other and may have to be repeated several times to achieve the desired performance.

$$R_1 = \frac{Q}{2\pi f_0 K c\, C_1} \qquad R_2 = K R_1 \qquad R_3 = \frac{1}{2\pi f_0 C_1}$$

The input impedance of op amps A_1 and A_3 should be $10 R_3$ or greater, and the input impedance of A_2 should be $10 R_1$ or greater.

This circuit should be used only for stages in which $Q \le 100$.

Fig. 6-26. Design formulas for second-order biquad bandpass filter stages corresponding to first-order lowpass stages.

Compute values for d and e as given by

$$e = \sqrt{\frac{c + 4Q^2 + \sqrt{(c + 4Q^2)^2 - (2bQ)^2}}{2b^2}}$$

$$d = \frac{be}{2Q} + \sqrt{\left(\frac{be}{2Q}\right)^2 - 1}$$

Fig. 6-27. Formulas for Design Procedure 6-8.

Design Procedure 6-8. To design a pair of second-order narrowband bandpass filter stages corresponding to a single second-order lowpass stage, perform the following steps:

1. Using formulas, tables or programs from Chapter 4 or 5, determine the value of the coefficients b and c for the transfer function of the lowpass stage corresponding to the desired pair of second-order bandpass stages.

2. Determine the desired stage gain K, quality factor Q, and center frequency f_o.

3. Pick a readily available standard value for C_1 that is close to $10^{-5}/f_o$.

4. Compute the coefficients d and e using the formulas shown in Fig. 6-27.

5. Determine the remaining capacitor values (if any), and compute the resistor values using the design formulas in Fig. 6-28 (VCVS), Fig. 6-29 (MFB), or Fig. 6-30 (biquad) as appropriate.

6. Select an op amp which has an input Z_{in} and an open loop gain A_{vo} meeting the conditions given in the appropriate set of design formulas.

7. Construct a pair of circuits in accordance with Fig. 6-21, 6-22, or 6-23 using the component values determined in Steps 4 and 5.

8. If necessary, tune the center frequency f_o, quality factor Q, and stage gain K by adjusting the resistor values as indicated in each schematic. These adjustments can interact with each other and may have to be repeated several times to achieve the desired performance.

Compute the A-stage resistor values as follows

$$R_{1A} = \frac{Q}{\pi f_0 K C_1 \sqrt{c}}$$

$$R_{2A} = \frac{-eQ}{\pi f_0 C_1 \left[dQ - \sqrt{(Ke\sqrt{c} - dQ)^2 + 8d^2 e^2 Q^2} \right]}$$

$$R_{3A} = \frac{R_{1A} + R_{2A}}{(2\pi f_0)^2 d^2 c^2 R_{1A} R_{2A}}$$

$$R_{4A} = 2R_{3A}$$

Compute the B-stage resistor values as follows

$$R_{1B} = \frac{Q}{\pi f_0 K C_1 \sqrt{c}}$$

$$R_{2B} = \frac{-deQ}{\pi f_0 C_1 \left[Q - \sqrt{(Kde\sqrt{c} - Q)^2 + e^2 Q^2} \right]}$$

$$R_{3B} = \frac{d^2(R_{1B} + R_{2B})}{(2\pi f_0)^2 C_1^2 R_{1B} R_{2B}}$$

$$R_{4B} = 2R_{3B}$$

The input impedance of the op amps used should be equal to or greater than $10R_{3A}$ or $10R_{3B}$ as appropriate.

This circuit should be used only for stages in which $Q \le 10$ and $KQ \le 100$.

Fig. 6-28. Design procedure for a pair of second-order VCVS bandpass filter stages corresponding to a single second-order lowpass stage.

Pick the smallest readily available standard value for C_2 such that

$$C_2 > \frac{K c^2}{Q^2} C_1$$

Compute the A-stage resistor values as follows

$$R_{1A} = \frac{Q}{2\pi f_0 K C_1 \sqrt{c}}$$

$$R_{2A} = \frac{Q}{2\pi f_0 \left[Q\, de\, (C_1 + C_2) - K C_1 \sqrt{c} \right]}$$

$$R_{3A} = \frac{e(C_1 + C_2)}{2\pi f_0 d C_1 C_2}$$

Compute the B-stage resistor values as follows

$$R_{1B} = \frac{Q}{2\pi f_0 K C_1 \sqrt{c}}$$

$$R_{2B} = \frac{Q d}{2\pi f_0 \left[Q\, e\, (C_1 + C_2) - K d C_1 \sqrt{c} \right]}$$

$$R_{3B} = \frac{de(C_1 + C_2)}{2\pi f_0 C_1 C_2}$$

The input impedance of the op amps used should be equal to or greater than $10 R_{3A}$ or $10 R_{3B}$ as appropriate.

This circuit should be used only for stages in which $Q \leq 10$ and $KQ \leq 100$.

Fig. 6-29. Design formulas for a pair of second-order MFB bandpass filter stages corresponding to a single second-order lowpass stage.

Compute the A-stage resistor values as follows

$$R_{1A} = \frac{Q}{2\pi f_0 K C_1 \sqrt{c}}$$

$$R_{2A} = \frac{K e \sqrt{c}}{d Q} R_{1A}$$

$$R_{3A} = \frac{1}{2\pi f_0 d C_1}$$

Compute the B-stage resistor values as follows

$$R_{1B} = \frac{Q}{2\pi f_0 K C_1 \sqrt{c}}$$

$$R_{2B} = \frac{K e d \sqrt{c}}{Q} R_{1B}$$

$$R_{3B} = \frac{d}{2\pi f_0 C_1}$$

The input impedance of op amps A_1 and A_2 should be equal to or greater than $10 R_{3A}$ or $10 R_{3B}$ as appropriate, and the input impedance of A_3 should be equal to or greater than $10 R_{1A}$ or $10 R_{1B}$

This circuit should be used only for stages in which $Q \leq 100$.

Fig. 6-30. Design formulas for a pair of second-order biquad bandpass filter stages correspond-ing to a single second-order lowpass stage.

Chapter 7

Fundamentals of
Digital Processing

D IGITAL SIGNAL PROCESSING IS BASED ON THE FACT THAT
an analog signal can be digitized and input to a general purpose digital computer or special purpose digital processor. Once
this is accomplished, we are free to perform all sorts of mathematical operations on the sequence of digital data samples inside the
processor. Some of these operations are simply digital versions of
classical analog techniques, while others have no counterpart in analog circuit devices or processing methods. In this chapter I will discuss the fundamentals of digitization and the various types of
processing that can be performed on the sequence of digital values
once they are inside the processor. Basically, you can chose between
two different types of approaches. You can adapt analog filter techniques to operate directly on the digital input sequence; or you can
compute the spectrum of the input, operate on this spectrum, and
then transform the result back into a time sequence. Digital transform methods will be covered in Chapter 9, and digital filters in
Chapter 8. In this chapter I will discuss some basic concepts involved in both approaches.

7.1 IDEAL SAMPLING

Inside a processor's memory, a digital signal consists of a sequence of separate individual values. It is useful and convenient to assume that each of these values represents the amplitude of an analog signal sampled at a single instant in time. The theoretical process of creating such a sequence of zero-width samples from the amplitude of a continuous analog signal is called "ideal sampling." As shown in Fig. 7-1, a mathematical model of an ideally sampled signal can be obtained by multiplying the original analog signal by a periodic train of unit impulses. In the output of a practical analog-to-digital converter (ADC) used to sample an actual signal, each sample value will of course exist for some nonzero interval of time. However, within the software of the digital processor, these values can still be interpreted as the amplitudes for a sequence of *ideal* samples. In fact, this is almost always the best approach since the ideal sampling model results in the simplest processing for most applications.

7.1.1 Spectral Changes Caused by Sampling. As established in Chapter 1, a periodic train of unit impulses in time will have a spectrum that consists of a periodic train of impulses in frequency. An ideally sampled signal can be viewed as the product of the original analog signal and a train of ideal impulses. Applying the signal multiplication property of the Fourier transform, you can quickly determine that the spectrum of the sampled-data signal is equal to the convolution of the original signal's spectrum with the spectrum of the impulse train. (Convolution was discussed in connection with the system impulse response in Chapter 2.) As shown in Fig. 7-2, this will result in a new spectrum consisting of copies of the original spectrum periodically replicated along the fre-

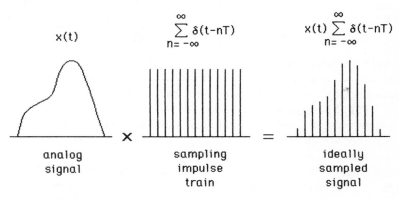

Fig. 7-1. Ideal sampling.

Spectrum of analog signal:

$$X(f) = \mathbf{F}[x(t)]$$

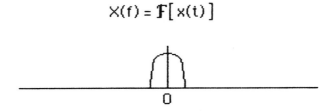

Spectrum of sampling impulse train:

$$\sum_{m=-\infty}^{\infty} \delta(f - mf_s)$$

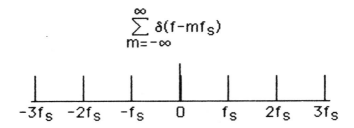

Spectrum of ideally sampled signal:

$$X(f) * \left(\sum_{m=-\infty}^{\infty} \delta(f - mf_s) \right) = \mathbf{F}\left[x(t) \sum_{n=-\infty}^{\infty} \delta(t - nT) \right]$$

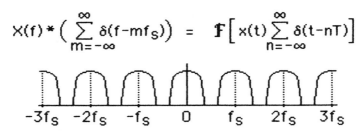

Fig. 7-2. Spectral changes caused by ideal sampling.

quency axis with period equal to the sampling rate f_s.

7.1.2 Sampling Rates. In order to perform effective sampling, you need to establish minimum usable sampling rates for the signals involved in any particular application. Clearly if you're trying to digitize human speech, one sample per second will be totally inadequate. On the other hand, 20 million samples per second will be outrageously fast and prohibitively expensive to accomplish.

You can establish a lower bound on the sampling rate for any spectrum of ideally sampled signal $(f_S > 2f_H)$

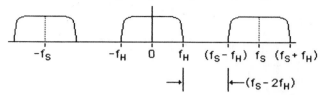

Fig. 7-3. Band spacing in the spectrum of an ideally sampled signal.

particular signal by examining the spectrum of an ideally sampled version of the signal. As shown in Fig. 7-3, the center-to-center spacing of the spectral bands is equal to the sampling rate f_s, while the edge-to-edge spacing of these strictly limited bands is equal to $f_s - 2f_H$. As long as f_s is greater than $2f_H$, the spectrum of the original signal can be recovered via a lowpass filtering operation that removes the replicated spectral bands introduced by the sampling process. This fact is formally stated by the uniform sampling theorem in Fig. 7-4, and is fundamental to the success of digital signal processing methods.

If f_s is less than $2f_H$, the individual bands of the spectrum will overlap as shown in Fig. 7-5, and lowpass recovery of the original signal will not be possible. This overlapping due to sampling at rates below $2f_H$ is called *aliasing*, while the minimum alias-free sampling rates of $2f_H$ is the *Nyquist rate*.

If the spectrum $X(f)$ of a function $x(t)$ vanishes beyond an upper frequency of f_H Hertz or ω_H radians per second, then $x(t)$ can be completely determined by its values at uniform intervals of less than $1/(2f_H)$ or π/ω_H. If sampled within these constraints, the original function $x(t)$ can be reconstructed from the samples by:

$$x(t) = \sum_{n=-\infty}^{\infty} \left[x(nT) \frac{\sin(2 f_s (t-nT))}{2 f_s (t-nT)} \right]$$

where T is the sampling interval.

Fig. 7-4. Uniform sampling theorem.

spectrum of ideally sampled signal $(f_S < 2f_H)$

(overlapping areas are shaded)

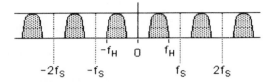

Fig. 7-5. Overlap of spectral bands due to undersampling.

7.2 DISCRETE-TIME SIGNALS

In the previous sections, weighted impulses were used to represent individual sample values in a discrete-time signal. This was necessary in order to use continuous mathematics to connect continuous-time analog signal representations with their corresponding discrete-time digital representations. However, once you are operating strictly within the digital or discrete-time realms, you can dispense with the Dirac delta impulse and adopt in its place the unit sample function which is much easier to work with. The unit sample function is also referred to as a *Kronecker delta impulse.* Figure 7-6 shows both the Dirac delta and Kronecker delta representations for a typical signal. In the equation for the Dirac impulse train, the independent variable is continuous time, t, and integer multiples of the sampling interval, T, are used to explicitly define the discrete sampling instants. On the other hand, the Kronecker delta notation assumes uniform sampling with an implicitly defined sampling interval. The independent variable is the integer-valued index n whose values correspond to the discrete instants at which samples can occur. The position of a specific sample is identified by the value of k which indicates the number of sampling intervals the Kronecker delta has been shifted from the origin at n = O. All of the common continuous-time functions can be sampled to produce their discrete-time counterparts.

7.3 DISCRETE-TIME SYSTEMS

In previous chapters you have seen how continuous-time systems such as filters and amplifiers can accept analog input signals and operate upon them to produce different analog output signals. *Discrete-time systems* perform essentially the same role for digital or discrete-time signals.

continuous signal

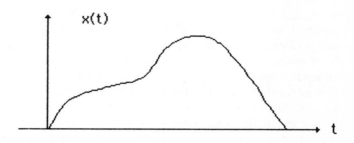

sampling with Dirac impulses

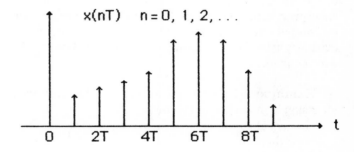

sampling with Kronecker impulses

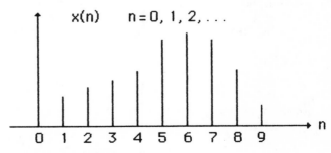

Fig. 7-6. Sampling with Dirac and Kronecker impulses.

7.3.1 Difference Equations. Although I have deliberately avoided discussing differential equations and their accompanying headaches in the analysis of analog systems, *difference* equations are much easier to work with and play an important role in the analysis of discrete-time systems. A discrete-time, linear, time-invariant (or if you prefer, shift-invariant) (DTLTI) system, which accepts an input sequence x(n) and produces an output sequence y(n), can be described by a linear difference equation of the form shown in Fig. 7-7. Such a difference equation can describe a DTLTI system having any initial conditions, as long as they are specified. This is in contrast to the discrete convolution and discrete transfer function which are limited to describing digital filters that are initially relaxed (i.e., all inputs and outputs are initially zero). In general, the computation of the output y(n) at point n using Equation 7.3-1 will involve previous outputs y(n − 1), y(n − 2), y(n − 3), etc. However in some filters, all of the coefficients a_1, a_2 . . . a_k are equal to zero as in Equation 7.3-2, and computation of y(n) does not involve previous output values. Difference equations involving previous output values are called *recursive* difference equations, and equations in the form of Equation 7.3-2 are called *nonrecursive* difference equations.

Example 7-1. Determine a nonrecursive difference equation for a simple moving-average lowpass filter in which the output at n = i is equal to the arithmetic average of the five inputs from n = (i-4) through n = i.

Solution. The desired difference equation is given by Equation 7.3-3 in Fig. 7-8. Relating this to the standard form of Equation 7.3-2, then k = 4, b_i = 0 for all i, and a_o = a_1 = a_2 = a_3 = a_4 = 0.2.

$$y(n) + a_1 y(n-1) + a_2 y(n-2) + \ldots + a_k y(n-k)$$

$$= b_0 x(n) + b_1 x(n-1) + b_2 x(n-2) + \ldots + b_k x(n-k)$$

(Eq. 7.3-1)

$$y(n) = b_0 x(n) + b_1 x(n-1) + b_2 x(n-2) + \ldots + b_k x(n-k)$$

(Eq. 7.3-2)

Fig. 7-7. Linear difference equations for describing discrete-time systems.

$$y(n) = \frac{x(n) + x(n-1) + x(n-2) + x(n-3) + x(n-4)}{5}$$

(Eq. 7.3-3)

$$y(n) = 0.2\,x(n) + 0.2\,x(n-1) + 0.2\,x(n-2) +$$

$$0.2\,x(n-3) + 0.2\,x(n-4)$$

(Eq. 7.3-4)

$$k = 4$$

$$b_i = 0 \quad \text{for all } i$$

$$a_0 = a_1 = a_2 = a_3 = a_4 = 0.2$$

$$a_i = 0 \quad \text{for } i > 4$$

Fig. 7-8. Equations for Example 7-1.

7.3.2 Discrete Convolution. A discrete-time system's *impulse response* is the output response produced when a unit sample function is applied to the input of the previously relaxed system. As you might suspect from our experiences with continuous systems, you can obtain the output $y(n)$ due to any input by performing a *discrete convolution* of the input signal $x(n)$ and the impulse response $h(n)$. This discrete convolution is given by Equation 7.3-5 in Fig. 7-9. If the impulse response $h(n)$ has nonzero values at an infinite number of points along the n-axis, the filter is called an *infinite impulse response* (IIR) filter. On the other hand, if $h(n) = 0$ for all $n \geq N$, the filter is called a *finite impulse response* (FIR) filter. FIR filters are also called *transversal* filters.

Example 7-2. For the moving-average filter described in Example 7-1, obtain the filter's impulse response.

Solution. The filter's impulse response $h(t)$ can be obtained by direct evaluation of Equation 7.3-3 for the case of $x(n)$ equal to the unit sample function as shown in Fig. 7-10.

$$y(n) = \sum_{m=0}^{n} x(m)\,h(n-m) \qquad\qquad \text{(Eq. 7.3-5)}$$

$$y(1) = x(0)\,h(1) + x(1)\,h(0)$$

$$y(2) = x(0)\,h(2) + x(1)\,h(1) + x(2)\,h(0)$$

$$\vdots$$

$$y(3) = x(0)\,h(n) + x(1)\,h(n-1) + \ldots + x(n)\,h(0)$$

Fig. 7-9. Discrete convolution.

$$y(n) = \frac{x(n) + x(n-1) + x(n-2) + x(n-3) + x(n-4)}{5}$$

$$\text{(Eq. 7.3-3)}$$

$$y(n) = h(n) \quad \text{when } x(n) = 0 \text{ for } n \neq 0$$

$$= 1 \text{ for } n = 0$$

for $n < 0$ $h(n) = 0.2(0 + 0 + 0 + 0 + 0) = 0$

$h(0) = 0.2(1 + 0 + 0 + 0 + 0)$

$h(1) = 0.2(0 + 1 + 0 + 0 + 0)$

$h(2) = 0.2(0 + 0 + 1 + 0 + 0)$

$h(3) = 0.2(0 + 0 + 0 + 1 + 0)$

$h(4) = 0.2(0 + 0 + 0 + 0 + 1)$

for $n > 0$ $h(n) = 0.2(0 + 0 + 0 + 0 + 0) = 0$

Fig. 7-10. Calculations for Example 7-2.

184

7.4 Z TRANSFORM TECHNIQUES

As discussed previously, a sampled-data signal can be represented as a series of Dirac impulse functions. Thus, the Laplace transform and all of the powerful analysis techniques associated with it could be applied to any sampled-data system whose impulse response is represented in this way. This all seems fairly straightforward, but there is one major difficulty associated with using Laplace transform techniques to analyze sampled-data systems. As shown in Section 7.1.1, sampling of a signal produces a

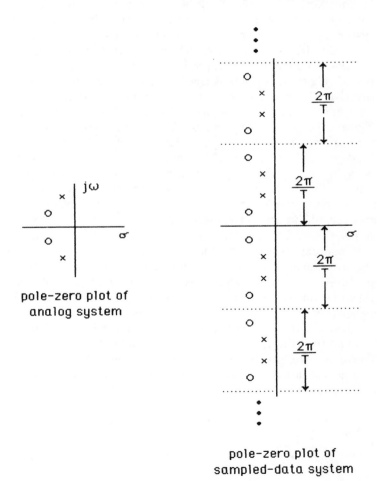

pole-zero plot of
analog system

pole-zero plot of
sampled-data system

Fig. 7-11. S-plane poles and zeros of a sampled-data system.

new spectrum consisting of copies of the original spectrum periodically replicated along the frequency axis. Although established using the Fourier transform and its properties, this condition is not limited to just the spectra of signals—a similar phenomenon is exhibited in the Laplace transforms of discrete-time systems. Figure 7-11 shows how sampling a continuous-time system will cause its pole-zero pattern to be replicated at intervals of f_s along the $j\omega$ axis. This confusing multitude of replicated poles and zeros can be avoided if you let $e^{sT} = z$ and thus define the Z *transform* given by Equation 7.4-3. For each pole of a continuous-time system there will be just one Z-transform pole in the corresponding discrete-time system. A number of well known transform pairs and transform properties are listed in Tables 7-1 and 7-2, respectively.

7.4.1 Discrete Transfer Functions. The transfer function of a digital filter can be derived from the linear difference equation that describes the filter. If you take the Z transform of each term in Equation 7.3-1, you obtain Equation 7.4-1. Factoring out $Y(z)$ and $X(z)$ and then solving for $Y(z)$ yields Equation 7.4-2. Both the numerator and denominator of $H(z)$ can then be factored to yield Equation 7.4-3. The poles of $H(z)$ are $p_1, p_2, \ldots p_k$, and the zeros are $q_1, q_2, \ldots q_m$. These poles and zeros can be plotted on the z-plane similar to the way in which poles and zeros of continuous-time systems are plotted in the s-plane, but there is a major difference regarding the pole locations of a stable system. In the s-plane, all poles of a stable system will lie to the left of the $j\omega$-axis; i.e., each real exponential component of the impulse response has a negative exponent and therefore converges as time increases. Since we set $e^{sT} = z$ and then plot the real and imaginary parts of z, the poles of stable systems must be confined within a unit circle in the z-plane.

Just like the inverse Laplace transform, the inverse Z transform involves the calculus of complex variables and direct evaluation of the inversion integral is rarely performed in actual practice. In practical situations, inversion of the Z transform is usually per-

$$X(z) = \sum_{n=0}^{\infty} x(n) z^{-n} \qquad \text{(Eq. 7.4-3)}$$

$$z = e^{sT}$$

Fig. 7-12. The Z transform.

Table 7-1. Common Z Transform Pairs.

$x(n)$	$X(z)$
1	$\dfrac{z}{z-1}$
$u_1(n)$	$\dfrac{z}{z-1}$
$\delta(n)$	1
nT	$\dfrac{Tz}{(z-1)^2}$
$\sin \omega_0 nT$	$\dfrac{z \sin \omega_0 T}{z^2 - 2z \cos \omega_0 T + 1}$
$\cos \omega_0 nT$	$\dfrac{z^2 - z \cos \omega_0 T}{z^2 - 2z \cos \omega_0 T + 1}$
e^{-anT}	$\dfrac{z}{z - e^{-aT}}$
$e^{-anT} \sin \omega_0 nT$	$\dfrac{z e^{-aT} \sin \omega_0 T}{z^2 - 2ze^{-aT} \cos \omega_0 T + e^{-2aT}}$
$e^{-anT} \cos \omega_0 nT$	$\dfrac{z^2 - ze^{-aT} \cos \omega_0 T}{z^2 - 2ze^{-aT} \cos \omega_0 T + e^{-2aT}}$

formed indirectly, using the transform pairs and transform properties listed in Tables 7-1 and 7-2. In addition, the partial-fraction expansion technique can be used to reduce complicated functions of z into a number of simpler functions that can then be found in Table 7-1.

7.5 PARTIAL FRACTION EXPANSION

A discrete transfer function of the general form given by Equa-

Table 7-2. Properties of the Z Transform.

Time function	Transform
$x(n)$	$X(z))$
$a\,x(n)$	$a\,X(z)$
$x(n) + y(n)$	$X(z) + Y(z)$
$e^{-anT}\,x(n)$	$X(e^{at}\,z)$
$x(n-m)$	$z^{-m}\,X(z)$
$\displaystyle\sum_{n=0}^{m} x(n)\,h(m-n)$	$X(z)\,H(z)$

tion 7.5-1 can be expanded into a sum of simpler terms that can be more easily inverse transformed. Linearity of the Z transform allows us to then sum the simpler inverse transforms to obtain the inverse of the original transfer function. The method for generating the expansion differs slightly depending on whether the transfer function's poles are all distinct or if some are multiple poles. Since most practical filter designs involve transfer functions with distinct poles, the more complicated multiple-pole procedure will not be presented. For a discussion of the multiple pole case, see *Cadzow, 1973.*

Design Procedure 7-2. To obtain the inverse Z transform of a discrete transfer function having all distinct poles, perform the following procedure. (Refer to Fig. 7-14 for equations.)

1. Factor the denominator of Equation 7.5-1 to produce Equation 7.5-2.
2. Compute c_0 as given by Equation 7.5-3.
3. Compute c_i for $i = 1$ through m using Equation 7.5-4.
4. Formulate the discrete-time function h(n) as given by Equation 7.5-5. The function h(n) is the inverse Z transform of H(z).

$$y(n) + a_1 y(n-1) + a_2 y(n-2) + \ldots + a_k y(n-k)$$

$$= b_0 x(n) + b_1 x(n-1) + b_2 x(n-2) + \ldots + b_k x(n-k)$$

(Eq. 7.3-1)

$$Y(z) + a_1 z^{-1} Y(z) + a_2 z^{-2} Y(z) + \ldots + a_k z^{-k} Y(z)$$

$$= b_0 X(z) + b_1 z^{-1} X(z) + b_2 z^{-2} X(z) + b_k z^{-k} X(z)$$

(Eq. 7.4-1)

$$Y(z) = \frac{b_0 + b_1 z^{-1} + b_2 z^{-2} + \ldots + b_k z^{-k}}{1 + a_1 z^{-1} + a_2 z^{-2} + \ldots + a_k z^{-k}} \; X(z)$$

(Eq. 7.4-2)

$$= H(z) \; X(z)$$

$$H(z) = \frac{b_0 (z-q_1)(z-q_2) \ldots (z-q_k)}{(z-p_1)(z-p_2)(z-p_3) \ldots (z-p_k)}$$

(Eq. 7.4-3)

Fig. 7-13. Discrete transfer function derived from a system's linear difference equation.

Example 7-3. Determine the inverse transform of Equation 7.5-6 shown in Fig. 7-15.

Solution. Comparing Equation 7.5-6 to the general form of Equation 7.5-1 reveals that $m = 2$, $b_o = 1$, $b_1 = 0$, and $b_2 = 0$. Factoring the denominator as in Equation 7.5-7 yields $p_1 = 1$ and $p_2 = -2$. Evaluation of c_0, c_1, and c_2 is straightforward, resulting in $h(n)$ as given by Equation 7.5-10.

$$H(z) = \frac{b_0 z^m + b_1 z^{m-1} + \ldots + b_{m-1} z^1 + b_m}{z_m + a_1 z^{m-1} + \ldots + a_{m-1} z^1 + a_m} \qquad \text{(Eq. 7.5-1)}$$

$$H(z) = \frac{b_0 z^m + b_1 z^{m-1} + \ldots + b_{m-1} z^1 + b_m}{(z-p_1)(z-p_2)(z-p_3)\ldots(z-p_k)} \qquad \text{(Eq. 7.5-2)}$$

$$c_0 = H(z)\Big|_{z=0} = \frac{b_m}{(-p_1)(-p_2)(-p_3)\ldots(-p_m)} \qquad \text{(Eq. 7.5-3)}$$

$$c_1 = \frac{z - p_i}{z} H(z)\Big|_{z=p_i} \qquad \text{(Eq. 7.5-4)}$$

$$h(n) = c_0 \delta(n) + c_1(p_1)^n + c_2(p_2)^n + \ldots + c_m(p_m)^n$$

$$\text{for } n = 0, 1, 2, \ldots \qquad \text{(Eq. 7.5-5)}$$

Fig. 7-14. Equations for Design Procedure 7-2.

$$H(z) = \frac{z^2}{z^2 + z - 2} \qquad \text{(Eq. 7.5-6)}$$

$$= \frac{z^2}{(z - 1)(z + 2)} \qquad \text{(Eq. 7.5-7)}$$

$$c_0 = 0$$

$$c_1 = \left(\frac{(z - 1)}{z} \frac{z^2}{(z - 1)(z + 2)} \right) \Bigg|_{z = 1} = \frac{z^2}{z^2 + 2z} \Bigg|_{z = 1} = \frac{1}{3}$$

$$\text{(Eq. 7.5-8)}$$

$$c_2 = \left(\frac{(z + 2)}{z} \frac{z^2}{(z - 1)(z + 2)} \right) \Bigg|_{z = -2} = \frac{z^2}{z^2 - z} \Bigg|_{z = -2} = \frac{2}{3}$$

$$\text{(Eq. 7.5-9)}$$

$$h(n) = \frac{1}{3}(1)^n + \frac{1}{3}(-2)^n$$

$$\text{(Eq. 7.5-10)}$$

$$= 1 + 2(-2)^n \quad 3 \qquad n = 0, 1, 2, \ldots$$

Fig. 7-15. Equations for Example 7-3.

Chapter 8

Digital Filters

D IGITAL FILTERS ARE USUALLY CLASSIFIED BY THE DURA-
tion of their impulse response, which can be either finite or
infinite. The methods for designing and implementing these two
filter classes differ considerably. *Finite impulse response* (FIR) filters
are digital filters whose response to a unit impulse is finite in duration. This is in contrast to *infinite impulse response* (IIR) filters whose
response to a unit impulse is infinite in duration. FIR and IIR filters
each have advantages and disadvantages, and neither is best in all
situations. Both FIR and IIR filters can be implemented using either recursive or nonrecursive techniques, but usually IIR filters
are implemented recursively and FIR filters are implemented via
nonrecursive methods.

8.1 ADVANTAGES AND DISADVANTAGES OF FIR FILTERS

FIR filters have the following advantages:

• FIR filters can easily be designed to have constant phase delay and constant group delay.
• FIR filters implemented with nonrecursive techniques will
always be stable and free from limit-cycle oscillations that often
plague IIR designs.
• Roundoff noise (which is due to finite precision arithmetic

performed in the digital processor) can be made relatively small for nonrecursive implementations.

• FIR filters can also be implemented using recursive techniques if this is desired.

Despite these advantages, FIR filters still exhibit some significant disadvantages:

• An FIR filter's impulse response duration, although finite, may have to be very long to obtain sharp cutoff characteristics.
• The design of FIR filters to meet specific performance objectives is generally more difficult than the design of IIR filters for similar applications.

8.2 FIR FILTER DESIGN

One way to generate a finite duration impulse response is to take an otherwise acceptable infinite duration impulse response and truncate it to a finite number of samples. This is equivalent to multiplying the infinite impulse response by a rectangular pulse or *window* that has a finite width of N samples. From the multiplication property of the FFT, you can show that this multiplication in the time domain will cause the frequency response of the IIR filter to be convolved with the frequency spectrum of the rectangular window, thus introducing ripples in the spectrum. This effect is closely related to both the Gibbs phenomenon and the leakage effect discussed in Section 9-5. This ripple can be greatly reduced by using a nonrectangular tapering window that has a narrower and more compact spectrum than a rectangular window. A number of useful windows and their spectra are presented at the conclusion of this chapter.

Other methods exist for design of FIR filters that are in some sense "optimal." However, these methods are generally more difficult to understand and implement than the window techniques presented here.

Design Procedure 8-1. To design an FIR filter using windows perform the following procedure.

1. Select the ideal analog filter which has the required passband and stopband configuration.
2. Choose a value N for the number of samples to be included

in the impulse response. This is sometimes referred to as the number of *taps* in an FIR filter. In general this number can be odd or even, but an odd number is usually preferred.

3. Determine the impulse response h(t) for the selected ideal analog filter. Shift h(t) so that it is symmetric about t = 0, and compute values for h(nT), n = 0, 1, 2, . . . (N – 1)/2—assuming of course that N is odd.

4. Multiply the samples of the impulse response h(nT) by one of the window sequences presented in the last section of this chapter. (Note that both the impulse response and window function will be symmetric around n = 0.) The resulting sequence is the impulse response of the desired FIR filter, which can be implemented using the techniques of Chapter 7.

8.3 REALIZATION OF FIR FILTERS

Equation 8.3-1 is the discrete convolution for an FIR filter. Conceptually, this can be implemented as shown in Fig. 8-1. However, actual construction of Fig. 8-1 for an N-stage filter would require

$$y(m) = \sum_{n=0}^{N-1} a_n \, x(m-n) \qquad (\text{Eq. } 8.3\text{-}1)$$

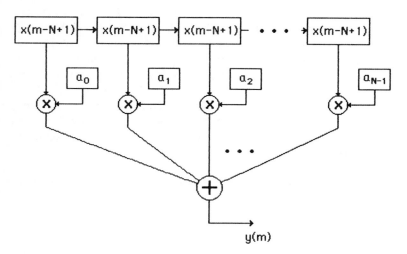

Fig. 8-1. Direct form realization of an FIR filter.

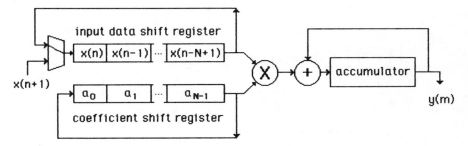

Fig. 8-2. Sequential structure for implementing FIR filters.

an N-stage data shifting element, N digital multipliers, N coefficient registers, and one N-input adder. Much more practical implementations can be designed to use significantly less hardware while performing the same mathematical operations. Figure 8-2 shows an FIR structure that is equivalent to Fig. 8-1, but which requires only two N-stage data shifting elements, one digital multiplier, and one two-input adder. The details of how the data and coefficients are shifted through this structure are illustrated in Fig. 8-3, which depicts the computation of y(3) and y(4) for a four-tap FIR filter.

8.4 IIR FILTER DESIGN
VIA THE BILINEAR TRANSFORMATION

It is possible to convert the transfer function of an analog filter into the transfer function of a digital filter by substituting $(z - 1)/(z + 1)$ for s. This substitution is called the *bilinear transformation* and is the basis for a popular method for designing IIR filters whose frequency responses approximate the responses of analog filters.

Design Procedure 8.2. To design an IIR lowpass filter using the bilinear transformation, perform the following procedure.

1. Determine the desired passband cutoff frequency Ω_c and stopband edge frequency Ω_1.
2. Compute "prewarped" analog frequencies ω_c and ω_1 using Equations 8.4-1 and 8.4-2.
3. Determine the transfer function H(s) for an analog lowpass filter having a cutoff frequency of ω_c and stopband edge of ω_1. This can be accomplished in a number of ways, including the methods presented in Chapters 4 and 5.

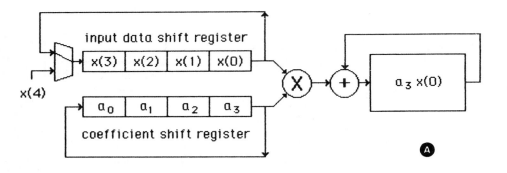

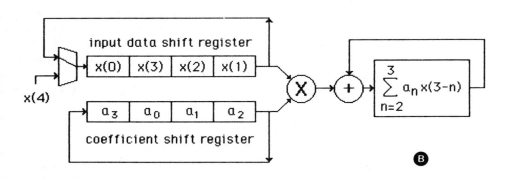

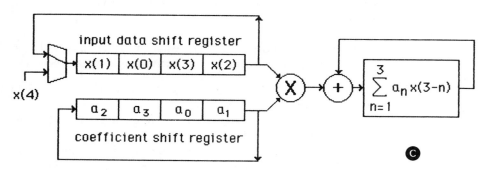

Fig. 8-3. Steps involved in computing y(3) and y(4) for a four-stage FIR filter. Continues through page 198.

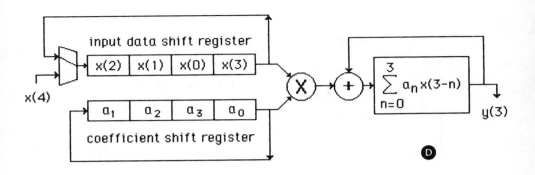

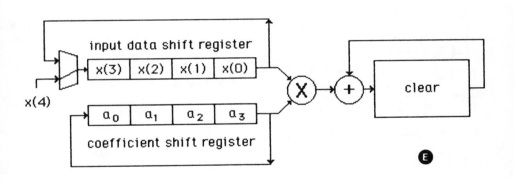

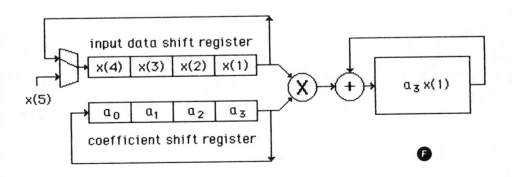

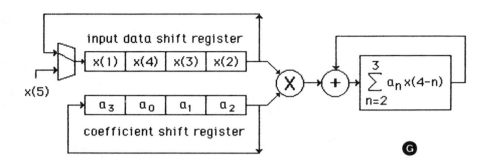

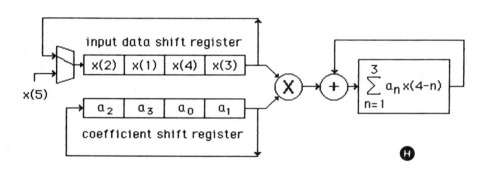

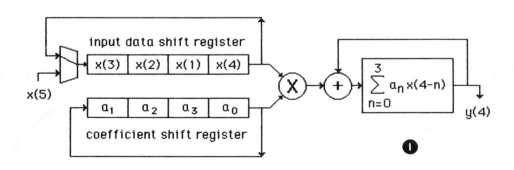

198

$$\omega_c = \tan \frac{\Omega_c T}{2} \qquad \text{(Eq. 8.4-1)}$$

$$\omega_1 = \tan \frac{\Omega_1 T}{2} \qquad \text{(Eq. 8.4-2)}$$

$$H(z) = H(s)\Big|_{s = \frac{z-1}{z+1}} \qquad \text{(Eq. 8.4-3)}$$

Fig.8-4. Formulas for Design Procedure 8-2.

4. Substitute $(z-1)/(z+1)$ for each occurrence of s in H(s) to form H(z), which is the transfer function of the desired digital low-pass filter.

8.5 REALIZATION OF IIR FILTERS

Realization of IIR filters can often be simplified if the discrete transfer function, H(z), is first converted into a linear difference equation.

Design Procedure 8-3. To obtain a linear difference equation from a system's discrete transfer function, perform the following steps:

1. Perform the algebra to put H(z) into the form of Equation 8.5-1.

2. Multiply this form of H(z) by X(z) to obtain Y(z) as in Equation 8.5-2. This can then be expanded to the form of Equation 8.5-3.

3. Using the Z-transform property given by Equation 8.5-4, take the inverse Z-transform of each term in Equation 8.5-3 to produce the linear difference equation given by 8.5-5, which can be readily implemented in either software or hardware.

WINDOW FUNCTIONS AND THEIR SPECTRA

Three of the more popular window functions—triangular, Hanning, and Hamming—are shown in Figs. 8A-2, 8A-4, and 8A-6 respectively. These were plotted using the BASIC program shown in Listing 8A-1. The spectra of these windows along with the spectrum of a rectangular window are shown in Figs. 8A-3, 8A-5, 8A-7, and 8A-1 respectively. These were plotted using the BASIC program shown in Listing 8A-2.

$$H(z) = \frac{b_0 + b_1 z^{-1} + b_2 z^{-2} + \ldots + b_k z^{-k}}{1 + a_1 z^{-1} + a_2 z^{-2} + \ldots + a_k z^{-k}} \qquad \text{(Eq.8.5-1)}$$

$$Y(z) = \frac{b_0 + b_1 z^{-1} + b_2 z^{-2} + \ldots + b_k z^{-k}}{1 + a_1 z^{-1} + a_2 z^{-2} + \ldots + a_k z^{-k}} \; X(z)$$

$$\text{(Eq. 8.5-2)}$$

$$Y(z) + a_1 z^{-1} Y(z) + a_2 z^{-2} Y(z) + \ldots + a_k z^{-k} Y(z)$$

$$= b_0 X(z) + b_1 z^{-1} X(z) + b_2 z^{-2} X(z) + b_k z^{-k} X(z)$$

$$\text{(Eq. 8.5-3)}$$

$$Z\left[x(n-m) \right] = z^{-m} X(z) \qquad \text{(Eq. 8.5-4)}$$

$$y(n) + a_1 y(n-1) + a_2 y(n-2) + \ldots + a_k y(n-k)$$

$$= b_0 x(n) + b_1 x(n-1) + b_2 x(n-2) + \ldots + b_k x(n-k)$$

$$\text{(Eq 8.5-5)}$$

Fig. 8-5. Equations for Design Procedure 8-3.

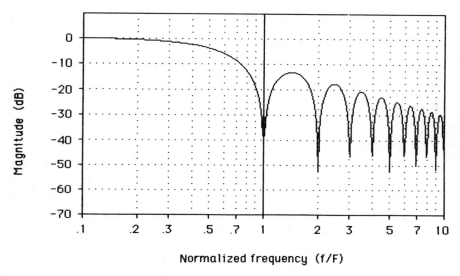

Fig. 8A-1. Magnitude spectrum of rectangular window.

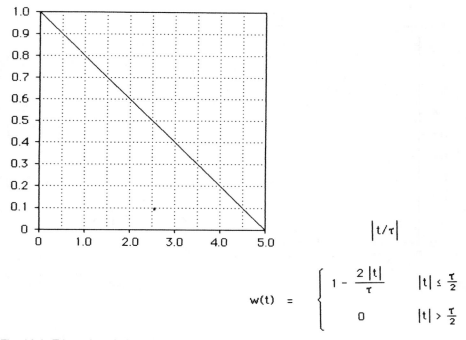

$$w(t) \ = \ \begin{cases} 1 - \dfrac{2\,|t|}{\tau} & |t| \leq \dfrac{\tau}{2} \\[2ex] 0 & |t| > \dfrac{\tau}{2} \end{cases}$$

Fig. 8A-2. Triangular window.

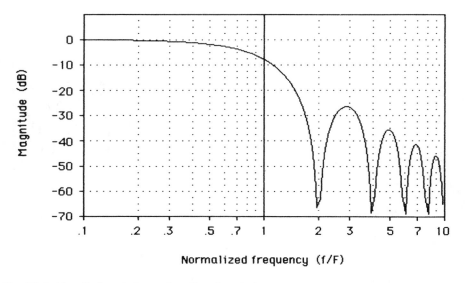

Fig. 8A-3. Magnitude spectrum of rectangular window.

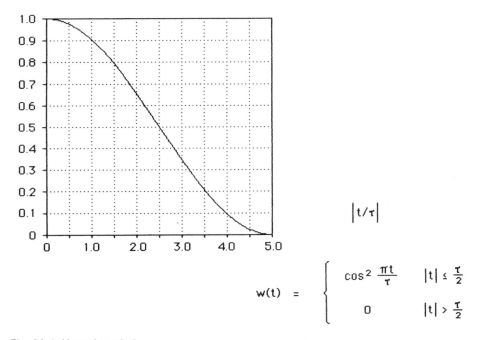

$$\left|t/\tau\right|$$

$$w(t) \;=\; \begin{cases} \cos^2 \dfrac{\pi t}{\tau} & |t| \leq \dfrac{\tau}{2} \\[2ex] 0 & |t| > \dfrac{\tau}{2} \end{cases}$$

Fig. 8A-4. Hamming window.

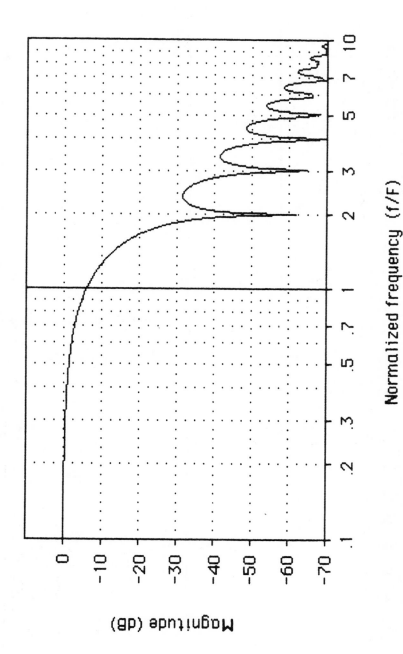

Fig. 8A-5. Magnitude spectrum of Hamming window.

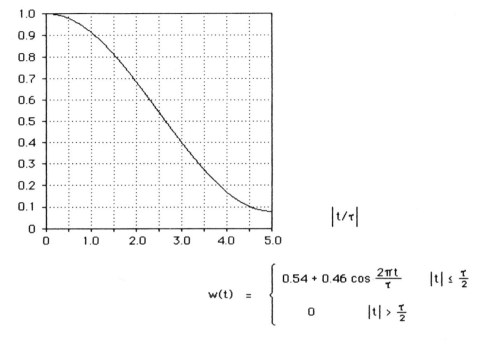

$$w(t) = \begin{cases} 0.54 + 0.46 \cos \frac{2\pi t}{\tau} & |t| \leq \frac{\tau}{2} \\ 0 & |t| > \frac{\tau}{2} \end{cases}$$

Fig. 8A-6. Hamming window.

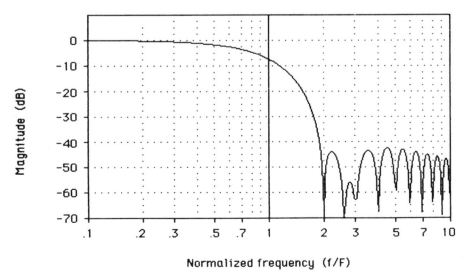

Fig. 8A-7. Magnitude spectrum of Hamming window.

204

Listing 8A-1. Program for plotting various window functions.

```
'******************************************************************
OPTION BASE 0
DIM yval(384)

INPUT "Alarm when done";alarm$
alarm$=UCASE$(alarm$)
INPUT "delta X";deltax%
CLS
PRINT "2  triangular window"
PRINT "3  Hanning window"
PRINT "4  Hamming window"
INPUT "Enter number of desired plot";jobtype%
CLS
IF jobtype%=2 THEN PRINT "triangular window"
IF jobtype%=3 THEN PRINT "Hanning window"
IF jobtype%=4 THEN PRINT "Hamming window"
CALL box.10.div.square
'
IF jobtype%=2 THEN CALL triangular.window(yval(),deltax%)
IF jobtype%=3 THEN CALL Hanning.window(yval(),deltax%)
IF jobtype%=4 THEN CALL Hamming.window(yval(),deltax%)
'
CALL window.plot(yval(),deltax%)
'
CALL dynamic.halt(alarm$)
'
'******************************************************************
'    SUBROUTINES
'******************************************************************
'    The following set of subprograms plots a square grid,
'    ten cycles by ten cycles
'
SUB box.10.div.square STATIC
FOR i%=0 TO 10
CALL vert.grid.line(50+i%*24)
CALL horiz.grid.line(20+i%*24)
NEXT i%
LINE (50,20)-(50,260)
```

```
LINE (50,20)-(290,20)
LINE (290,20)-(290,260)
LINE (50,260)-(290,260)
END SUB
.

.
SUB vert.grid.line(ix%) STATIC
FOR iy%=20 TO 260 STEP 4
PSET (ix%,iy%)
NEXT iy%
LINE (ix%,260)-(ix%,263)
END SUB
.

.
SUB horiz.grid.line(iy%) STATIC
FOR ix%=50 TO 290 STEP 4
PSET (ix%,iy%)
NEXT ix%
LINE (46,iy%)-(50,iy%)
END SUB
.
'*********************************************************
'   The following subprogram computes a triangular window function
.
SUB triangular.window(yval(1),deltax%) STATIC
FOR ix%=0 TO 239
yval(ix%)=1!-ix%/240!
NEXT ix%
END SUB
.
'*********************************************************
'   The following subprogram computes a Hanning window function
.
SUB Hanning.window(yval(1),deltax%) STATIC
pi#=3.1415926535898#
.
FOR ix%=0 TO 239
t#=(ix%/240#)/2#
yval(ix%)=(COS(pi#*t#))^2
NEXT ix%
```

```
END SUB
'
'*******************************************************
'   The following subprogram computes a Hamming window function
'
SUB Hamming.window(yval(1),deltax%) STATIC
pi#=3.1415926535898#

FOR ix%=0 TO 239
t#=(ix%/240#)/2#
yval(ix%)=.54+.46*COS(2*pi#*t#)
NEXT ix%
'
END SUB
'
'*******************************************************
'   The following subprogram plots the window function contained
'   in the vector yval()
'
SUB window.plot(yval(1),deltax%) STATIC
ixold%=0
iyold%=260!-240!*yval(0)
FOR ix%=1 TO 239 STEP deltax%
iy%=260!-240!*yval(ix%)
LINE (ixold%+50,iyold%)-(ix%+50,iy%)
iyold%=iy%
ixold%=ix%
NEXT ix%
END SUB
'*******************************************************
SUB dynamic.halt(alarm$) STATIC
done.loop:
IF alarm$<>"Y" THEN GOTO done.loop
SOUND 1000,10,250
SOUND 700,10,250
GOTO done.loop
END SUB
```

Listing 8A-2. Program for plotting sprectra of various window functions.

```
'**********************************************************
'
OPTION BASE 0
DIM yval(384)
'
e# = 2.7182818284592#
pi#=3.1415926535898#
'
INPUT "Alarm when done";alarm$
alarm$=UCASE$(alarm$)
INPUT "delta X";deltax%
CLS
PRINT "1  spectrum of rectangular window"
PRINT "2  spectrum of triangular window"
PRINT "3  spectrum of Hanning window"
PRINT "4  spectrum of Hamming window"
INPUT "Enter number of desired plot";jobtype%
CLS
IF jobtype%=1 THEN PRINT "spectrum of rectangular window"
IF jobtype%=2 THEN PRINT "spectrum of triangular window"
IF jobtype%=3 THEN PRINT "spectrum of Hanning window"
IF jobtype%=4 THEN PRINT "spectrum of Hamming window"
CALL two.cyc.semi.log.box
'
IF jobtype%=1 THEN CALL rectangular.spectrum(yval(),deltax%,10,2,1!)
IF jobtype%=2 THEN CALL triangular.spectrum(yval(),deltax%,10,2,1!)
IF jobtype%=3 THEN CALL Hanning.spectrum(yval(),deltax%,10,2,1!)
IF jobtype%=4 THEN CALL Hamming.spectrum(yval(),deltax%,10,2,1!)
'
CALL response.plot(yval(),deltax%,-70!,10!,1)
'
CALL dynamic.halt(alarm$)
'
'**********************************************
'   SUBROUTINES
'**********************************************
'  The following set of subprograms plots a semi-logarithmic grid,
'  two cycles by eight divisions
```

```
SUB two.cyc.semi.log.box STATIC
CALL box.384.by.224
CALL two.cycle.freq.grid(7)
FOR iy%=48 TO 216 STEP 28
CALL horizontal.grid.line(iy%,7)
NEXT iy%
END SUB
'

SUB box.384.by.224 STATIC
LINE (50,20)-(50,248)
LINE (46,20)-(438,20)
LINE(46,244)-(438,244)
LINE(434,20)-(434,248)
END SUB
'

SUB two.cycle.freq.grid(dot.interval%) STATIC
DIM two.cycle.freq.tics%(17)
two.cycle.freq.tics%(0)=108
two.cycle.freq.tics%(1)=141
two.cycle.freq.tics%(2)=166
two.cycle.freq.tics%(3)=184
two.cycle.freq.tics%(4)=200
two.cycle.freq.tics%(5)=213
two.cycle.freq.tics%(6)=224
two.cycle.freq.tics%(7)=233
two.cycle.freq.tics%(8)=242
two.cycle.freq.tics%(9)=300
two.cycle.freq.tics%(10)=333
two.cycle.freq.tics%(11)=358
two.cycle.freq.tics%(12)=376
two.cycle.freq.tics%(13)=392
two.cycle.freq.tics%(14)=405
two.cycle.freq.tics%(15)=416
two.cycle.freq.tics%(16)=425
FOR tic.index%=0 TO 16
ix%=two.cycle.freq.tics%(tic.index%)
FOR iy%=20 TO 244 STEP dot.interval%
IF (iy%-20) MOD 28 <>0 THEN PSET(ix%,iy%)
NEXT iy%
```

```
LINE(ix%,244)-(ix%,248)
NEXT tic.index%
LINE(242,20)-(242,244)
END SUB
'
SUB horizontal.grid.line(iy%,dot.interval%) STATIC
LINE(46,iy%)-(50,iy%)
FOR ix%=50 TO 434 STEP dot.interval%
PSET (ix%,iy%)
NEXT ix%
LINE(434,iy%)-(438,iy%)
END SUB
'
'***********************************************************
'
'   The following subprogram computes the rectangular window's spectrum.
'
SUB rectangular.spectrum(yval(1),deltax%,maxfreq%,freqcyc%,tau) STATIC
e# = 2.7182818284592#
m# = 1
pi#=3.1415926535898#
maxfreqexp%=LOG(maxfreq%)/2.302585093#
'
'*******************
'   Compute normalizing factor equal to dc response of window
'
ix%=0
f# = (10^(maxfreqexp%+freqcyc%*(ix%-384)/384))
x#=pi#*f#*tau
'
sinc#=0
FOR n%=0 TO 40
'
factor#=1!/x#
FOR j%=1 TO 2*n%+1
factor#=factor#*(x#/j%)
NEXT j%
sinc#=sinc#+((-1)^n%)*factor#
NEXT n%
'
```

210

```
norm.fact#=tau*ABS(sinc#)
yval(ix%)=20!*LOG(norm.fact#)/2.302585093#
'
'*******************
'   Compute window response at all other frequencies
'
FOR ix%=1 TO 383 STEP deltax%
CALL MOVETO(350,13)
PRINT 383-ix%
f# = (10^(maxfreqexp%+freqcyc%*(ix%-384)/384))
x#=pi#*f#*tau
'
sinc#=0
FOR n%=0 TO 40
'
factor#=1!/x#
FOR j%=1 TO 2*n%+1
factor#=factor#*(x#/j%)
NEXT j%
sinc#=sinc#+((-1)^n%)*factor#
NEXT n%
'
yval(ix%)=20!*LOG(tau*ABS(sinc#)/norm.fact#)/2.302585093#
'
NEXT ix%
'
CALL MOVETO(350,13)
PRINT"     "
END SUB
'
'***********************************************************
'
'   The following subprogram computes the triangular window's spectrum.
'
SUB triangular.spectrum(yval(1),deltax%,maxfreq%,freqcyc%,tau) STATIC
e# = 2.7182818284592#
m# = 1
pi#=3.1415926535898#
maxfreqexp%=LOG(maxfreq%)/2.302585093#
```

```
'*********************
' Compute normalizing factor equal to dc response of filter
'
ix%=0
f# = (10^(maxfreqexp%+freqcyc%*(ix%-384)/384))
x#=pi#*f#*tau/2!
'
sinc#=0
FOR n%=0 TO 40
'
factor#=1!/x#
FOR j%=1 TO 2*n%+1
factor#=factor#*(x#/j%)
NEXT j%
sinc#=sinc#+((-1)^n%)*factor#
NEXT n%
'
norm.fact#=(tau/2!)*sinc#*sinc#
'PRINT norm.fact#
yval(0)=20!*LOG(1!)/2.302585093#
'
'*******************
' Compute window response for all other frequencies
'
FOR ix%=1 TO 383 STEP deltax%
CALL MOVETO(350,13)
PRINT 383-ix%
'
f# = (10^(maxfreqexp%+freqcyc%*(ix%-384)/384))
x#=pi#*f#*tau/2!
'
sinc#=0
FOR n%=0 TO 40
'
factor#=1!/x#
FOR j%=1 TO 2*n%+1
factor#=factor#*(x#/j%)
NEXT j%
sinc#=sinc#+((-1)^n%)*factor#
NEXT n%
```

212

```
'
yval(ix%)=20!*LOG((tau/2!)*sinc#*sinc#/norm.fact#)/2.302585093#
'
NEXT ix%
'
CALL MOVETO(350,13)
PRINT"    "
END SUB
'
'*************************************************************
'
'  The following subprogram computes the Hanning window's spectrum.
'
SUB Hanning.spectrum(yval(1),deltax%,maxfreq%,freqcyc%,tau) STATIC
e# = 2.7182818284592#
m# = 1
pi#=3.1415926535898#
maxfreqexp%=LOG(maxfreq%)/2.302585093#
'
'*******************
'  Compute normalizing factor equal to dc response of window
'
ix%=0
f# = (10^(maxfreqexp%+freqcyc%*(ix%-384)/384))
x#=pi#*f#*tau
'
sinc#=0
FOR n%=0 TO 40
'
factor#=1!/x#
FOR j%=1 TO 2*n%+1
factor#=factor#*(x#/j%)
NEXT j%
sinc#=sinc#+((-1)^n%)*factor#
NEXT n%
f.tau.sqrd#=(f#*tau)^2
norm.fact#=(tau/2!)*ABS(sinc#*(1!/(1!-f.tau.sqrd#)))
yval(0)=20!*LOG(1!)/2.302585093#
'
'*******************
```

```
'   Compute window response at all other frequencies
'

FOR ix%=1 TO 383 STEP deltax%
CALL MOVETO(350,13)
PRINT 383-ix%
f# = (10^(maxfreqexp%+freqcyc%*(ix%-384)/384))
x#=pi#*f#*tau
'
sinc#=0
FOR n%=0 TO 40
'
factor#=1!/x#
FOR j%=1 TO 2*n%+1
factor#=factor#*(x#/j%)
NEXT j%
sinc#=sinc#+((-1)^n%)*factor#
NEXT n%
'
f.tau.sqrd#=(f#*tau)^2
linear.response#= (tau/2!)*ABS(sinc#*(1!/(1!-f.tau.sqrd#)))/norm.fact#
yval(ix%)=20!*LOG(linear.response#)/2.302585093#
'
NEXT ix%
'
CALL MOVETO(350,13)
PRINT"      "
END SUB
'
'*********************************************************
'
'   The following subprogram computes the Hamming window's spectrum.
'
SUB Hamming.spectrum(yval(1),deltax%,maxfreq%,freqcyc%,tau) STATIC
e# = 2.7182818284592#
m# = 1
pi#=3.1415926535898#
maxfreqexp%=LOG(maxfreq%)/2.302585093#
'
'********************
'   Compute normalizing factor equal to dc response of window
```

214

```basic
ix%=0
f# = (10^(maxfreqexp%+freqcyc%*(ix%-384)/384))
x#=pi#*f#*tau

sinc#=0
FOR n%=0 TO 40

factor#=1!/x#
FOR j%=1 TO 2*n%+1
factor#=factor#*(x#/j%)
NEXT j%
sinc#=sinc#+((-1)^n%)*factor#
NEXT n%

f.tau.sq#=(f#*tau)^2
norm.fact#=tau*sinc#*((.54-.08*f.tau.sq#)/(1!-f.tau.sq#))
yval(0)=20!*LOG(1!)/2.302585093#

'***********************
'   Compute window response at all other frequencies
'
FOR ix%=1 TO 383 STEP deltax%
CALL MOVETO(350,13)
PRINT 383-ix%
f# = (10^(maxfreqexp%+freqcyc%*(ix%-384)/384))
x#=pi#*f#*tau

sinc#=0
FOR n%=0 TO 40

factor#=1!/x#
FOR j%=1 TO 2*n%+1
factor#=factor#*(x#/j%)
NEXT j%
sinc#=sinc#+((-1)^n%)*factor#
NEXT n%

f.tau.sq#=(f#*tau)^2
lin.resp#= tau*ABS(sinc#*((.54-.08*f.tau.sq#)/(1!-f.tau.sq#)))/norm.fact#
```

215

```
yval(ix%)=20!*LOG(lin.resp#)/2.302585093#
'
NEXT ix%
'
CALL MOVETO(350,13)
PRINT"      "
END SUB
'
'********************************************************
'
'   The following subprogram plots the data contained in the vector yval()
'
SUB response.plot(yval(1),deltax%,ymin,ymax,tracetype%) STATIC
yrange=ymax-ymin
iyold%=244-224*((yval(0)-ymin)/yrange)
iyold2%=INT(244!-224!*((yval(0)-ymin)/yrange))
ixold%=0
FOR ix%=1 TO 383 STEP deltax%
iy%=244-224*((yval(ix%) - ymin)/yrange)
iy2%=INT(244!-224!*((yval(ix%)-ymin)/yrange))
IF iy%>244 THEN GOTO eof.pb.plot.loop
IF tracetype%=2 THEN PSET(ix%+50,iy%)
IF tracetype%=1 THEN LINE(ixold%+50,iyold%)-(ix%+50,iy%)
IF tracetype%=3 THEN LINE(ixold%+50,iyold2%)-(ix%+50,iy2%)
IF tracetype%=3 THEN LINE(ixold%+50,iyold2%+1)-(ix%+50,iy2%+1)
ixold%=ix%
iyold%=iy%
iyold2%=iy2%
eof.pb.plot.loop:
NEXT ix%
END SUB
'
'********************************************************
'
SUB dynamic.halt(alarm$) STATIC
done.loop:
IF alarm$<>"Y" THEN GOTO done.loop
SOUND 1000,10,250
SOUND 700,10,250
GOTO done.loop
END SUB
```

Chapter 9

Digital Transform Techniques

A S WE HAVE SEEN, SPECTRAL ANALYSIS CAN BE AN EX-tremely useful tool in processing signals. However, the analog spectral methods covered so far (Fourier transform, Laplace transform) are not well suited for direct use on small computer systems. There are some artificial intelligence type, problem-solving "expert systems" that can perform the calculus involved in the closed-form analysis of continuous signals and systems. However, such expert systems involve tremendous hardware and software resources to solve even the simplest problems. Furthermore, they require closed-form representations of the input signal and the processing operations to be performed. Often we need to process signals derived from instrumentation of real-world processes, and are unable to obtain the required closed-form expressions. In order to bring economical and effective spectral analysis into the small computer arena, the Fourier transform of Chapter 1 needs to be modified somehow so that it can operate on a discrete-time function to produce the appropriate discrete-frequency spectrum. This chapter will present this new transform and explore its limitations and practical application.

9.1 DISCRETE FOURIER TRANSFORM

The discrete Fourier transform (DFT) shown in Fig. 9-1 can

217

be derived from the continuous transform. Notice that all traces of calculus have disappeared and we now have a simple algebraic summation in place of the continuous transform's integral. The summations involved in both the DFT (Equations. 9.1-1 and 9.1-2) and inverse DFT (IDFT) (Equations. 9.1-3 and 9.1-4) each contain a finite number of terms and thus are ideally suited for direct evaluation on a digital processor.

Example 9-1. Write a BASIC subprogram to compute the DFT of a 32-point real-valued input sequence assuming that T = 9.25 msec and F = 5 Hz. Assume that the vector **Little.x[ ]** already contains the time sequence values. Put the real and imaginary parts of the resulting frequency sequence into the vectors **Big.X.real[ ]** and **Big.X.imag[ ]**, respectively.

Solution. The desired program is shown in Listing 9-1.

9.1.1 DFT Parameters. In Example 9-1, we were given values for N, T, and F, but in designing a DFT for a particular application, values must be chosen for these parameters. N is the number of time sequence values x(n) over which the DFT summation

$$X(m) = \sum_{n=0}^{N-1} x(n) e^{-j2\pi mnFT} \qquad m = 0, 1, \ldots N-1$$

$$(Eq. 9.1\text{-}1)$$

$$= \sum_{n=0}^{N-1} x(n) \cos(2\pi mnFT) + j \sum_{n=0}^{N-1} x(n) \sin(2\pi mnFT)$$

$$(Eq. 9.1\text{-}2)$$

$$x(n) = \sum_{m=0}^{N-1} X(m) e^{j2\pi mnFT} \qquad n = 0, 1, \ldots N-1$$

$$(Eq. 9.1\text{-}3)$$

$$= \sum_{m=0}^{N-1} X(m) \cos(2\pi mnFT) + j \sum_{m=0}^{N-1} X(m) \sin(2\pi mnFT)$$

$$(Eq. 9.1\text{-}4)$$

Fig. 9-1. Discrete Fourier transform.

is performed to compute each frequency sequence value. It is also the total number of frequency sequence values X(m) produced by the DFT. In discussions of DFT operation, a complete set of N consecutive time sequence values is often called an *input record,* and a complete set of N frequency sequence values is called an *output record.* T is the time interval between two consecutive samples of the time sequence, and F is the frequency interval between two consecutive samples of the frequency sequence. The selection of these values is subject to the following constraints, which are imposed by the sampling theorem (see Section 7.1) and the inherent properties of the DFT:

1. The assumptions implicit in the DFT require that $FNT = 1$.
2. To avoid aliasing, the sampling theorem requires that $T \leq 1/(2f_H)$, where f_H is the highest significant frequency component in the continuous-time signal.
3. The record length in time is equal to NT or 1/F.
4. Many "fast" DFT algorithms will require that N be an integer power of two or at least highly composite.

Example 9-2. Choose values of N, F, and T given that F must be 5 Hz or less, N must be an integer power of 2, and the bandwidth of the input signal is 300 Hz. For the values chosen, determine the longest signal that can fit into a single input record.

Solution. From constraint #2 above, $T \leq 1/(2f_h)$. Since $f_h = 300$ Hz, $T \leq 1.66$ msec. If you select $F = 5$ and $T = .0016$, then $N \geq 125$. Since N must be an integer power of 2 then choose $N = 128$. Using these values, the input record will span 204.8 msec.

Example 9-3. Assuming that $N = 256$ and F must be 5 Hz or less, determine the highest input signal bandwidth that can be accomodated.

Solution. Since $FNT = 1$, then $T \geq 781.25$ μsec. This corresponds to a maximum f_h of 640 Hz.

9.1.2 Periodicity. As discussed previously, a periodic function of time will have a discrete-frequency spectrum (see Section 1.4), and a discrete-time function will have a spectrum that is periodic (Section 7.2). Since the DFT relates a discrete-time function to a corresponding discrete-frequency function, this implies that both the time function and frequency function are periodic as well as discrete. (See Fig. 9-2.) This means that some care must be ex-

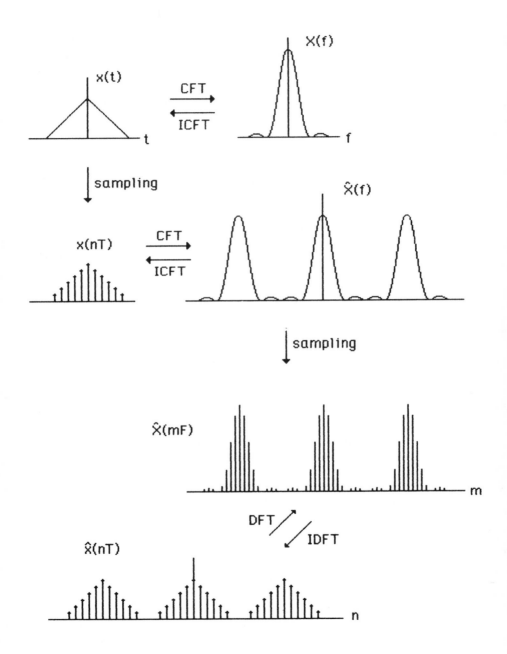

Fig. 9-2. Relationships between continuous and discrete-time functions and their spectra.

ercised in selecting DFT parameters and interpreting DFT results, but it does *not* mean that the DFT can only be used on periodic digital signals. Based on this inherent periodicity, it is a common practice to regard the points from n = 1 through n = N/2 as positive and the points from n = N/2 through n = N − 1 as negative. Since both the time and frequency sequences are periodic, the values at points n = N/2 through n = N − 1 are in fact equal to the values at points n = − N/2 through n = − 1. Under this convention, it is convenient to redefine the concept of even and odd sequences: If x(N − n) = x(n) then x(n) is even, and if x(N − n) = − x(n) then x(n) is odd.

9.2 PROPERTIES OF THE DFT

The DFT exhibits a number of useful properties and operational relationships that are similar to the properties of the continuous Fourier transform discussed in Chapter 1. The DFT versions of these properties are discussed in the sections below.

9.2.1 Linearity. The DFT is both homogeneous and

The DFT, relating x(n) and X(m) :

$$x(n) \quad \underset{IDFT}{\overset{DFT}{\rightleftarrows}} \quad X(m)$$

is homogeneous −

$$a\,X(n) \quad \underset{IDFT}{\overset{DFT}{\rightleftarrows}} \quad a\,X(m), \qquad \text{(Eq. 9.2-1)}$$

additive −

$$x(n) + y(n) \quad \underset{IDFT}{\overset{DFT}{\rightleftarrows}} \quad X(m) + Y(m), \qquad \text{(Eq. 9.2-2)}$$

and therefore linear −

$$a\,x(n) + b\,y(n) \quad \underset{IDFT}{\overset{DFT}{\rightleftarrows}} \quad a\,X(m) + b\,Y(m) \qquad \text{(Eq. 9.2-3)}$$

Fig. 9-3. Linear properties of the DFT.

Given

$$x(n) \xrightarrow[\text{IDFT}]{\text{DFT}} X(m)$$

then

$$\frac{1}{N} X(n) \xrightarrow[\text{IDFT}]{\text{DFT}} x(-m) \qquad (Eq.\ 9.2\text{-}4)$$

Fig. 9-4. Symmetric property of the DFT.

additive—and therefore linear as well. Mathematical descriptions of these three properties are shown in Fig. 9-3.

9.2.2 Symmetry. Equation 9.2-4 of Fig. 9-4 shows the symmetry that exists between a time sequence and the corresponding frequency sequence produced by the DFT.

9.2.3 Time Shifting. A time sequence x(n) can be delayed in time by subtracting an integer from n. Delaying the time sequence will cause the corresponding frequency sequence to be phase shifted as in Equation 9.2-5 of Fig. 9-5.

9.2.4 Frequency Shifting. Time sequence modulation is accomplished by multiplying the time sequence by an imaginary exponential term $e^{j2\pi nk/N}$. This will cause a frequency shift of the spectrum as in Equation 9.2-9 of Fig. 9-6.

9.2.5 Even and Odd Properties of the DFT. The DFT of an even time sequence is a frequency sequence which is both

Given

$$x(n) \xrightarrow[\text{IDFT}]{\text{DFT}} X(m)$$

then

$$x(n-k) \xrightarrow[\text{IDFT}]{\text{DFT}} X(m)\ e^{-j2\pi mk/N}$$

$$(Eq.\ 9.2\text{-}5)$$

Fig. 9-5. Time-shifting property of the DFT.

Given

$$x(n) \xrightleftharpoons[\text{IDFT}]{\text{DFT}} X(m)$$

then

$$x(n) \, e^{\, j2\pi mk/N} \xrightleftharpoons[\text{IDFT}]{\text{DFT}} X(m-k) \qquad (\text{Eq.} 9.2\text{-}6)$$

Fig. 9-6. Modulation or frequency-shifting property of the DFT.

real-valued and even, while the DFT of an odd time sequence is imaginary and odd as in Fig. 9-7.

9.2.6 Real and Imaginary Properties of the DFT. In general, the DFT of a real-valued time sequence will have an even real component and an odd imaginary component. Conversely, an

---

Consider a time sequence $x(n)$ and the corresponding frequency sequence $X(m) = X_r(m) + jX_i(m)$.

If $x(n)$ is even,

$$x(-n) = x(n)$$

then

$$X(m) = X_r(m) = X_r(-m) \qquad (\text{Eq.} 9.2\text{-}7)$$

($X(m)$ is real-valued and even.)

If $x(n)$ is odd,

$$x(-n) = -x(n)$$

then

$$X(m) = X_i(m) = -X_i(-m) \qquad (\text{Eq.} 9.2\text{-}8)$$

($X(m)$ is imaginary and odd.)

Fig. 9-7. Even and odd properties of the DFT.

imaginary-valued time sequence will have an odd real component and an even imaginary component, as shown in Fig. 9-8.

9.3 DFT OF SHORT TIMELIMITED SIGNALS

Consider the timelimited continuous-time signal and its continuous spectrum shown in Fig. 9-9(A and B). (Remember, a signal cannot be both strictly timelimited and strictly bandlimited.) We can sample this signal to produce the time sequence shown in Fig. 9-9C for input to a DFT. If the input record length, N, of the DFT is chosen to be longer than the length of the input time sequence, the entire sequence can fit within the input record as shown. As discussed in Section 9.1.2, the DFT will treat the input sequence as though it is the periodic sequence shown in Fig. 9-9D. This will result in a periodic discrete-frequency spectrum as shown in Fig. 9-9E. The actual output produced by the DFT algorithm will be the sequence of values from $m = 0$ to $m = N - 1$. Of course, there will be some aliasing due to the timelimited nature (and consequently unlimited bandwidth) of the input signal pulse. However,

---

Given a time sequence $x(n) = x_r(n) + jx_i(n)$ and the corresponding frequency sequence $X(m) = X_r(m) + jX_i(m)$:

If $x(n)$ is real

$$x(n) = x_r(n)$$

then

$$X_r(m) = X_r(-m) \qquad \text{(Eq. 9.2-9a)}$$

$$X_i(m) = -X_i(-m) \qquad \text{(Eq. 9.2-9b)}$$

If $x(n)$ is imaginary

$$x(n) = x_i(n)$$

then

$$X_r(m) = -X_r(-m) \qquad \text{(Eq. 9.2-10a)}$$

$$X_i(m) = X_i(-m) \qquad \text{(Eq. 9.2-10b)}$$

Fig. 9-8. Real and imaginary properties of the DFT.

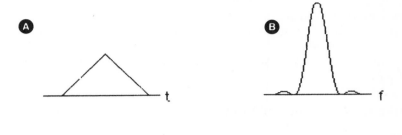

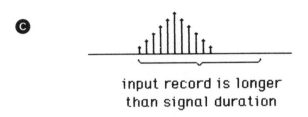

input record is longer
than signal duration

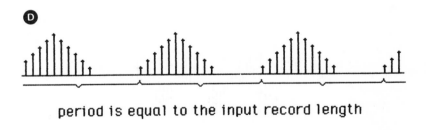

period is equal to the input record length

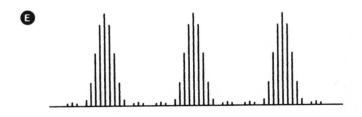

Fig. 9-9. Signals and sequences in the DFT of a short timelimited signal.

some applications do involve certain signal pulse shapes that are both reasonably timelimited and yet still bandlimited enough to avoid serious aliasing. Aside from these aliasing problems (which are actually attributable to the sampling operation), the DFT can be used quite effectively in the analysis of these types of timelimited signal pulses.

9.4 DFT OF PERIODIC SIGNALS

Consider the bandlimited and periodic continuous-time signal and its spectrum shown in Fig. 9-10. We can sample this signal to produce the time sequence shown in Fig. 9-10C for input to a DFT. If the input record length, N, of the DFT is chosen to be exactly equal to length of one period of this sequence, the periodic assumptions made by the DFT will cause the DFT to treat the sin-

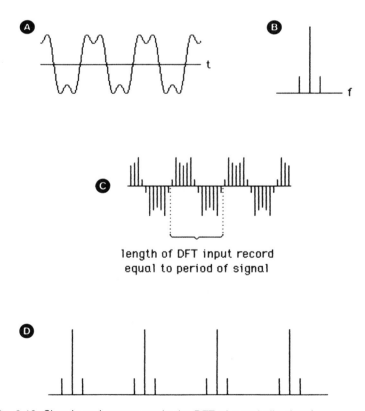

length of DFT input record
equal to period of signal

Fig. 9-10. Signals and sequences in the DFT of a periodic signal.

gle input record as though it is the complete sequence. The corresponding periodic discrete-frequency spectrum is shown in Fig. 9-10D. The DFT output sequence will actually consist of just one period which matches *exactly* the spectrum of Fig. 9-10B. You could not hope for (or find) a more convenient situation. Unfortunately, this relationship exists *only* in an N-point DFT where the input signal is both bandlimited and periodic with a period of exactly N.

9.5 DFT OF LONG APERIODIC SIGNALS

Sections 9.2 and 9.3 covered the use of the DFT under relatively favorable conditions that are not likely to exist in many important signal processing applications. Often the signal to be analyzed will be neither periodic nor reasonably timelimited. The corresponding sequence of digitized signal values will usually be longer than the DFT input record and will therefore have to be truncated to just N samples before the DFT can be applied. The peri-

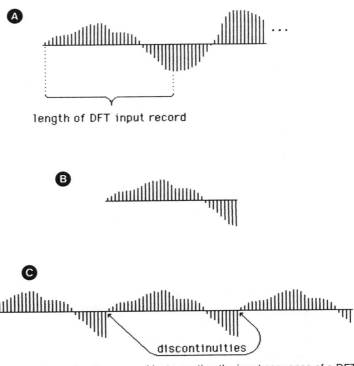

Fig. 9-11. Discontinuities caused by truncating the input sequence of a DFT.

227

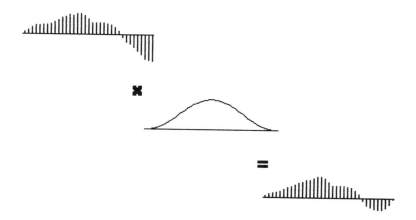

discontinuity is gone

Fig. 9-12. Use of a tapering window to reduce time-domain discontinuities.

odic nature of the DFT will cause the truncated sequence of Fig. 9-11B to be interpreted as though it is the sequence shown in Fig. 9-11C. Notice that in this sequence there is a large discontinuity in the signal at points corresponding to the ends of the input record. This will introduce additional high frequency components into the spectrum produced by the DFT. This phenomenon is called *leakage*. To reduce the leakage effects, it is a common practice to multiply the truncated input sequence by a tapering window as in Fig. 9-12 prior to application of the DFT. A good window shape will taper off at the ends of the input record but still have a reasonably compact and narrow spectrum. This is important since multiplying the time sequence by the window will cause the corresponding frequency sequence to be convolved with the spectrum of the window. A narrow window spectrum will cause minimum smearing of the signal spectrum due to this convolution. Several popular windowing functions and their spectra are shown at the end of Chapter 8.

9.6 DFT COMPUTATION

The DFT is a powerful signal analysis tool, but for large values of N it can require prohibitively large numbers of mathematical operations. Using the expanded DFT summation in Fig. 9-13, it can readily be seen that computation of X(m) for any value of m will require $N - 1$ multiplications and $N - 1$ additions. Therefore, computing a complete set of N values for X(m) will entail $(N - 1)^2$ multiplications and $N(N - 1)$ additions. (The multiplications for computing X(O) are trivial hence they are not included in this count.) Furthermore, you need to compute the values of $e^{-j2\pi mn/N}$ for various values of m and n. However, this is not usually a serious problem since $e^{-j2\pi mn/N}$ is periodic with a period of N. (Remember—$e^{-j2\pi mn/N} = \cos(2\pi mn/N) - j \sin(2\pi mn/N)$.) Therefore the relationships of Fig. 9-14 can be used to limit the number of different exponential values to be computed. An N point DFT will involve at most N unique values of $e^{-j2\pi mn/N}$. These can easily be computed once and then stored in a table for use as needed during the actual DFT computation.

In analysis and development of different DFT algorithms, the exponential term $e^{-j2\pi mn/N}$ appears so often that an alternative notation has been widely adopted. For any given size of DFT, the values of j, 2, π, and N are all constant, so it is common to define $W_N = e^{-j2\pi/N}$ thus making $e^{-j2\pi mn/N} = W_N^{mn}$. When the value of N is obvious from context, the subscript is often omitted.

In general, the multiplications and additions in the DFT can involve complex values, and in comparing the computational "cost" of various algorithms most writers assume that these operations *always* involve complex values. As shown in Figs. 9-15 and 9-16, the operations involve in DFT computations can actually be of six

$$X(m) = \sum_{n=0}^{N-1} x(n)\, e^{-j2\pi mn/N}$$

$$= x(0) + x(1)\, e^{-j2\pi m/N} + x(2)\, e^{-j4\pi m/N} + \dots$$

$$\dots + x(N-1)\, e^{-j2(N-1)\pi m/N} \tag{Eq. 9.6-1}$$

Fig. 9-13. Expansion of the DFT summation for one value in the frequency sequence.

$$e^{-2\pi j} = 1 \qquad\qquad e^{-\pi j/2} = j$$

$$e^{-\pi j} = -1 \qquad\qquad e^{-3\pi j/4} = -j$$

$$e^{-2k\pi j} = 1 \qquad (k = \text{integer})$$

$$e^{(-2\pi j/N)(kN + r)} = e^{-2\pi r j/N}$$

$$e^{-2\pi jk/N} = e^{-2\pi j(k \text{ MOD } N)/N}$$

Fig. 9-14. Properties useful for computing exponential terms in the DFT.

different types—real with real, real with imaginary, real with complex, imaginary with imaginary, imaginary with complex, and complex with complex. Each of these types involves different numbers of real operations. Since the cost of implementing any particular algorithm will depend on the number of real operations to be performed, you need to establish the specific types of operations involved and determine the equivalent number of real operations. For the general case of x(n) having complex values, the DFT will involve $(N-1)^2$ complex multiplications and $N(N-1)$ complex additions. In terms of real operations this corresponds to $4(N-1)^2$ multiplications and $(4N^2 - 6N + 1)$ additions. In most practical situations, however, the time sequence x(n) will be real-valued. Then the DFT will involve $(N-1)^2$ multiplications of a real value with a complex value (W^{mn} is still complex) and $N(N-1)$ complex additions. In terms of real operations this corresponds to $2(N-1)^2$ multiplications and $2(N^2 - N)$ additions. As N becomes large, the number of operations involved in a direct DFT increases proportional to N^2 and can become prohibitively large. Fortunately, a number of clever algorithms have been devised which drastically reduce the number of operations required to compute a DFT.

9.7 FAST FOURIER TRANSFORMS

The various algorithms that implement large-N DFTs with significantly fewer than N^2 complex multiplications are collectively referred to as *fast Fourier transforms* (FFTs). The rigorous derivation of general FFT forms requires the use of matrix factorization techniques that are somewhat abstract and difficult to follow. How-

Real times real

$$(a,0)\,(c,0) = (ac,0)$$ one real multiplication

Real times imaginary

$$(a,0)\,(0,dj) = (0,adj)$$ one real multiplication

Real times complex

$$(a,0)\,(c,dj) = (ac,adj)$$ two real multiplications

Imaginary times imaginary

$$(0,bj)\,(0,dj) = (-bd,0)$$ one real multiplication

Imaginary times complex

$$(0,bj)\,(c,dj) = (-bd,bcj)$$ two real multiplications

Complex times complex

$$(a,bj)\,(c,dj) = ((ac-bd),(ad+bc)j)$$ four real multiplications
two real additions

Fig. 9-15. Types of multiplication found in DFT algorithms.

ever, it is possible to understand and use FFT techniques without understanding all the rigors involved in their development. On the other hand, using "magical" techniques with no grasp of how they work the way they do is not particularly satisfying. Therefore, just to gain an appreciation of how these fast algorithms relate to the basic DFT, I will develop a specific algorithm for an eight-point transform. For all of the basic fast algorithms, the number of samples in both the time and frequency sequences must be an integer power of two—i.e., 4, 8, 16, 32, etc.

Real plus real

$$(a,0) + (c,0) = (a+c,0)$$ one real addition

Real plus imaginary

$$(a,0) + (0,dj) = (a,dj)$$ no operations

Real plus complex

$$(a,0) + (c,dj) = (a+c,dj)$$ one real addition

Imaginary plus imaginary

$$(0,bj) + (0,dj) = (0,(b+d)j)$$ one real addition

Imaginary plus complex

$$(0,bj) + (c,dj) = (c,(b+d)j)$$ one real addition

Complex plus complex

$$(a,bj) + (c,dj) = (a+c,(b+d)j)$$ two real additions

Fig. 9-16. Types of addition found in DFT algorithms.

232

$X(0) = x(0) \, W^0 + x(1) \, W^0 + x(2) \, W^0 \quad + x(3) \, W^0 \quad + x(4) \, W^0 \quad + x(5) \, W^0 \quad + x(6) \, W^0 \quad + x(7) \, W^0$

$X(1) = x(0) \, W^0 + x(1) \, W^1 + x(2) \, W^2 \quad + x(3) \, W^3 \quad + x(4) \, W^4 \quad + x(5) \, W^5 \quad + x(6) \, W^6 \quad + x(7) \, W^7$

$X(2) = x(0) \, W^0 + x(1) \, W^2 + x(2) \, W^4 \quad + x(3) \, W^6 \quad + x(4) \, W^8 \quad + x(5) \, W^{10} + x(6) \, W^{12} + x(7) \, W^{14}$

$X(3) = x(0) \, W^0 + x(1) \, W^3 + x(2) \, W^6 \quad + x(3) \, W^9 \quad + x(4) \, W^{12} + x(5) \, W^{15} + x(6) \, W^{18} + x(7) \, W^{21}$

$X(4) = x(0) \, W^0 + x(1) \, W^4 + x(2) \, W^8 \quad + x(3) \, W^{12} + x(4) \, W^{16} + x(5) \, W^{20} + x(6) \, W^{24} + x(7) \, W^{28}$

$X(5) = x(0) \, W^0 + x(1) \, W^5 + x(2) \, W^{10} + x(3) \, W^{15} + x(4) \, W^{20} + x(5) \, W^{25} + x(6) \, W^{30} + x(7) \, W^{35}$

$X(6) = x(0) \, W^0 + x(1) \, W^6 + x(2) \, W^{12} + x(3) \, W^{18} + x(4) \, W^{24} + x(5) \, W^{30} + x(6) \, W^{36} + x(7) \, W^{42}$

$X(7) = x(0) \, W^0 + x(1) \, W^7 + x(2) \, W^{14} + x(3) \, W^{21} + x(4) \, W^{28} + x(5) \, W^{35} + x(6) \, W^{42} + x(7) \, W^{49}$

Fig. 9-17. Computation of an eight-point DFT.

$$X(0) = ((x(0) + x(4) W^0) + W^0 (x(2) + x(6) W^0)) + W^0 ((x(1) + x(5) W^0) + W^0 (x(3) + x(7) W^0))$$

$$X(1) = ((x(0) + x(4) W^4) + W^2 (x(2) + x(6) W^4)) + W^1 ((x(1) + x(5) W^4) + W^2 (x(3) + x(7) W^4))$$

$$X(2) = ((x(0) + x(4) W^0) + W^4 (x(2) + x(6) W^0)) + W^2 ((x(1) + x(5) W^0) + W^4 (x(3) + x(7) W^0))$$

$$X(3) = ((x(0) + x(4) W^4) + W^6 (x(2) + x(6) W^4)) + W^3 ((x(1) + x(5) W^4) + W^6 (x(3) + x(7) W^4))$$

$$X(4) = ((x(0) + x(4) W^0) + W^0 (x(2) + x(6) W^0)) + W^4 ((x(1) + x(5) W^0) + W^0 (x(3) + x(7) W^0))$$

$$X(5) = ((x(0) + x(4) W^4) + W^2 (x(2) + x(6) W^4)) + W^5 ((x(1) + x(5) W^4) + W^2 (x(3) + x(7) W^4))$$

$$X(6) = ((x(0) + x(4) W^0) + W^4 (x(2) + x(6) W^0)) + W^6 ((x(1) + x(5) W^0) + W^4 (x(3) + x(7) W^0))$$

$$X(7) = ((x(0) + x(4) W^4) + W^6 (x(2) + x(6) W^4)) + W^7 ((x(1) + x(5) W^4) + W^6 (x(3) + x(7) W^4))$$

Fig. 9-18. Factored equations for computation of an eight-point DFT.

9.8 A FAST DFT ALGORITHM FOR N = 8

The computation of X(0) through X(7) for an eight-point (N = 8) DFT is shown in Fig. 9-17. Since W^{mn} is periodic with a period of N, $W^{mn} = W^{mn \pm kN}$ where k is any integer. Making use of this fact, equations in Fig. 9-17 can be factored to produce the equations shown in Fig. 9-18. Examination of these equations reveals that they share many common terms that can be computed once and then used as often as needed without having to be computed over again. This is the source of the computational savings offered by the fast algorithms. Use of these common terms is easier to understand if the equations are presented in the form of a signal flow graph as in Fig. 9-19. Such signal flow graphs are a compact way of representing various DFT algorithms. Each line represents a term or combination of terms from the factored DFT equations, with each circle representing one (possibly complex) addition and one (possibly complex) multiplication. The term represented by the

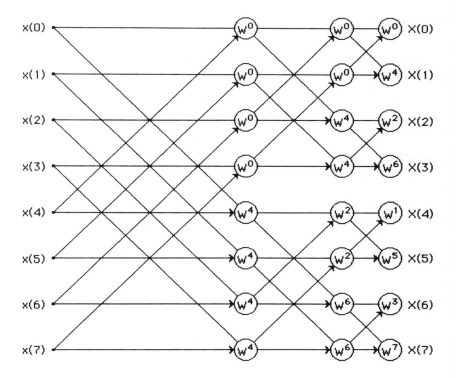

Fig. 9-19. Signal flow graph representing the equations of Fig. 9-18.

line with an arrowhead entering the circle is multiplied by the constant within the circle and added to the term represented by the other entering line. The result is then represented by the lines exiting the circle.

Since there are 24 circles in Fig. 9-19, this algorithm requires 24 multiplications and 24 additions compared to 64 multiplications and 64 additions for the direct DFT computation. In general, for an N-point transform, this algorithm will require $N \log_2 N$ multiplications and the same number of additions.

Listing 9-1. BASIC program for Example 9-1.

```
'*******************************************************

'    Listing 9-1

'    BASIC subprogram for Example 9-1

'*******************************************************

SUB DFT(Little.x(1), Big.X.real(1), Big.X.imag(1) ) STATIC

pi# = 3.1415926535898#
T = .00625
F = 5!

FOR m%=0 TO 31
Big.X.real(m%) = 0
Big.X.imag(m%) = 0

FOR n%=0 TO 31
Big.X.real(m%) = Big.X.real(m%) + Little.x(n%) * COS(2*pi#*m%*n%*F*T)
Big.X.imag(m%) = Big.X.imag(m%) + Little.x(n%) * SIN(2*pi#*m%*n%*F*T)
NEXT n%

NEXT m%
END SUB
```

Bibliography

Beachamp, K.G. and Yuen, C.K. *Digital Methods for Signal Analysis*, George Allen & Unwin LTD, 1979.

Stearns, S.D. *Digital Signal Analysis*, Hayden Book Co., 1975.

Gold, B. and Rader, C.M. *Digital Processing of Signals*, McGraw-Hill, 1969.

Peled, A. and Liu, B. *Digital Signal Processing*, Wiley, 1976.

Antoniou, A *Digital Filters: Analysis and Design*, McGraw-Hill, 1979.

Chen, Chi-Tsong *One-Dimensional Signal Processing*, Marcel Decker, 1979.

Otnes, R.K. and Enochson, L. *Applied Time Series Analysis*, Wiley, 1978.

Otnes, R.K. and Enochson, L. *Digital Time Series Analysis*, Wiley, 1972.

Lancaster, D. *Active-Filter Cookbook*, Howard W. Sams, 1975.

Rabiner, L.R. and Gold, B. *Theory and Application of Digital Signal Processing*, Prentice-Hall, 1975.

Bogner, R.E. and Constantinides, A.G. *Introduction to Digital Filtering*, Wiley.

Hsu, H.P. *Fourier Analysis*, Simon & Schuster, 1970.

Spiegel, M.R. *Laplace Transforms*, McGraw-Hill, 1965.

Brigham, E.O. *The Fast Fourier Transform*, Prentice-Hall, 1974.

Stanley, W.D. *Digital Signal Processing,* Reston Pub. Co., 1975.

Schwartz, R.J. and Friedland, B. *Linear Systems*, McGraw-Hill, 1965.

Lynn, P.A. *An Introduction to the Analysis and Processing of Signals*, Howard W. Sams, 1982.

Oppenheim, A.V. and Schafer, R.W. *Digital Signal Processing*, Prentice-Hall, 1975.

Cadzow, J.A. *Discrete-Time Systems*, Prentice-Hall, 1973.

Johnson, D.E., Johnson, J.R. and Moore, H.P. *A Handbook of Active Filters*, Prentice-Hall, 1980.

Williams, A.B. *Electronic Filter Design Handbook,* McGraw-Hill, 1981.

Blinchikoff, H.J. and Zverev, A.I. *Filtering in the Time and Frequency Domains*, Wiley, 1976.

Berlin, H.M. *Design of Active Filters with Experiments*, Howard W. Sams, 1977.

Tedeschi, F.P. *The Active Filter Handbook*, TAB Books Inc., 1979.

Nussbaumer, H.J. *Fast Fourier Transform and Convolution Algorithms*, Springer-Verlag, 1981.

Rabiner, L.R. and Rader, C.M., editors *Digital Signal Processing*, IEEE Press, 1972. (This is a collection of reprints of landmark papers in digital signal processing. The major areas covered are digital filters, FFTs, and the effects of finite word lengths in digital processing.)

Selected Papers in Digital Signal Processing II, edited by DSP committee of the IEEE ASSP Soc., IEEE Press, 1976.

Index

Index